THE HAM RADIO PREP

General Class License Manual

2020 - 2024

Copyright © 2022 by American Radio Club

All rights reserved. No part of this publication may be reproduced, distributed or transmitted in any form or by any means, including photocopying, recording, or other electronic or mechanical methods, without the prior written permission of the publisher, except in the case of brief quotations embodied in critical reviews and certain other noncommercial uses permitted by copyright law. For permission requests, write to the publisher, addressed "Attention: Permissions Coordinator," at the address below.

1309 Coffeen Ave Ste 1956
Sheridan, WY 82801

Or visit: HamRadioPrep.com

Ordering information:

Special discounts are available on quantity purchases by corporations, associations and others. For details, contact the publisher at the address below.

Email: contact@HamRadioPrep.com

Printed in the United States of America

ISBN: 9798746031518

Imprint: Independently published

This book is dedicated to Rachel. Thank you for being there every step of the way, without your support, this book would not have been possible.

DISCLAIMER

You must have an amateur radio license from the Federal Communications Commission to transmit on any amateur radio frequencies. Transmitting without a license can result in heavy fines and/or imprisonment. This book is intended to help you pass your license exam and is not meant to serve as an operator's manual. Ham Radio Prep or American Radio Club Inc. are not liable for your on-air operations. For correct operation please refer to Part 97 of the FCC rules.

HAM RADIO PREP
Operators Creed

Licensed amateur radio operators play an important role in our nation and world. They:

Operate on the air legally and follow Federal Communications Commission rules and side agreements that allow others to use the airwaves for furtherance of the hobby.

Use their radio equipment and stations in a way that benefits themselves as a hobby but also their community and nation, and even world. Amateur radio is a valuable resource that always must be put to good use for the benefit of all.

Utilize spectrum efficiently and properly so that amateur radio has a presence throughout the spectrum to ensure frequencies are available for generations of new amateurs to come.

Embrace emerging technologies and digital standards and platforms so that they are advancing the art of amateur radio by being on the air.

Respect fellow amateurs on the air and ensure that amateur radio frequencies are a welcoming place for all to gather, no matter their background.

Prepare others to learn the science and art of amateur radio so that we always are increasing our ranks to ensure that our spectrum is utilized as much as possible.

Use the hobby to learn not only more about science and technology, but also our family and neighbors.

Table of Contents

Chapter 1: Commission's Rules — 1

Chapter 2: Operating Procedures — 33

Chapter 3: Radio Wave Propagation — 61

Chapter 4: Amateur Radio Practices — 82

Chapter 5: Electrical Principles — 112

Chapter 6: Circuit Components — 147

Chapter 7: Practical Circuits — 165

Chapter 8: Signals and Emissions — 182

Chapter 9: Antennas and Feed lines — 202

Chapter 10: Electrical and RF Safety — 224

Practice Exams — 238

Scheduling and Taking Your Exam — 286

Helpful Resources — 301

Answer Keys — 309

Helpful links and Resources

Full Online Course:
https://hamradioprep.com

Free Lesson:
https://hamradioprep.com/free-lesson

Online Testing Information:
https://hamradioprep.com/ham-radio-license-test-online/

10 Hardest Questions on the Exam:
https://hamradioprep.com/hardest-questions-on-fcc-technician-license-test/

Scheduling an Exam:
https://hamradioprep.com/schedule-an-exam/

Why do you need a ham license?
https://hamradioprep.com/why-do-you-need-a-ham-radio-license/

How to Register for your FRN:
https://hamradioprep.com/how-to-register-for-your-frn/

Thank you and welcome!

HAM RADIO PREP

Thank you for purchasing the Ham Radio Prep Technician License manual. This book has been carefully constructed using years of development to help you pass your ham radio license exam easily.

We founded Ham Radio Prep in 2017, when a group of aspiring ham radio operators went to get our ham radio licenses. We found that all of the materials were too confusing, too long, too boring, and we were NEVER going to be able to get our licenses.

As it turns out, a lot of other people felt the same way. And this was preventing many Americans from taking the first step of getting into ham radio - which we believe is critical for independent communications, public service, STEM education, and more.

So, with almost no money in our pockets, we decided to create a new program to make getting your ham radio license FAST, EASY, and FUN!

It's been a lot of blood, sweat and tears since 2017 when we started the program, but we can now happily say, Ham Radio Prep is the most popular program in the world for studying for your ham radio license. We've had over 60,000 students use our program to pass their ham radio license tests. We've used all of the proceeds of every purchase to continue to improve our program and support our students.

As a token of our appreciation, we would like to share a coupon code with you for any of our online programs.

If you have already used our online program, feel free to share this with a friend, or many friends! Share the love and let's get more folks licensed and on the air.

www.HamRadioPrep.com

Of course, this book works great on its own - it has been carefully designed to cover the full question pool and you will pass the exam successfully.

However, the online course adds in video with animations, games, and a full interactive media experience.

This helps reinforce the concepts for different learning styles, and you also have extra benefits like unlimited practice tests. If you enjoy learning through video as well as a multimedia program, we encourage you to check it out!

You can even try a free lesson at www.HamradioPrep.com to see what our online program is all about.

Unlock **100%** of all amateur radio frequencies with an Extra class license!

Start studying and save **20%** on Amateur Extra or **ANY** of our amazing prep courses.

To Redeem:
1. Visit www.HamRadioPrep.com/pricing
2. At checkout, click on the "Have a coupon?" button and enter the following code
3. Click "Apply"

Your Coupon Code:

GENBOOK479

Need additional support?

See an error that needs to be corrected?

Please reach out to our student success team at support@hamradioprep.com for any assistance you may need at all!

Accuracy of material is crucial to the success of any learning program, and Ham Radio Prep is no exception. We take errors very seriously. One of our agents will reach out to you within one business day of when you submitted the ticket for any follow up necessary.

We greatly appreciate your help in making Ham Radio Prep the best it can be!

Study on the go!

Download our free quiz and practice exam simulator for more practice while on the go. Search for Ham Radio Prep in the Apple App Store and Google Play Store to get started. Both apps include practice for the Technician, General and Extra exams.

Our other social media:

www.**facebook**.com/HamRadioPrep
www.**twitter**.com/HamPrep
www.**instagram**.com/HamRadioPrep
www.**reddit**.com/user/HamRadioPrep
www.**youtube**.com/c/HamRadioPrep
www.**tiktok**.com/@HamRadioPrep
www.**linkedin**.com/c/HamRadioPrep

FCC General license exam breakdown by chapter

The actual FCC General license exam that you will take has a total of 35 questions, which are derived from 10 subpools of questions (which is why our manual is broken up into 10 chapters).

To create the practice exam correctly, we use the same number of questions from each subpool as they will use on your actual exam. This makes our practice tests reliable for testing your knowledge of the required questions.

Section	Number of questions	% of exam
Chapter 1: Commission's rules	5	14%
Chapter 2: Operating procedures	5	14%
Chapter 3: Radio wave propagation	3	9%
Chapter 4: Amateur radio practices	5	14%
Chapter 5: Electrical principles	3	9%
Chapter 6: Circuit components	2	5%
Chapter 7: Practical circuits	3	9%
Chapter 8: Signals and emissions	3	9%
Chapter 9: Antennas and feed lines	4	12%
Chapter 10: Electrical and RF safety	2	5%

How to use this book

Each chapter in this book was written to give context and understanding to the questions in the FCC General license exam pool.

At the end of each chapter, you will find a quiz to review the information you just learned. Use a pencil to mark the answer you think is correct.

Check your work using the answer key located in the back of the book. To calculate your quiz scores, multiple the number of questions you answered right by 10. If you answered seven questions correctly, you scored 70 percent.

At the end of Chapter 10, you will find three full-length practice exams. Our quizzes and practice exams use the exact same questions as on the official FCC exam. Although the actual questions may vary, the number of questions derived from each section follows the same requirements as the FCC exam.

Again, using a pencil to mark the answer you think is correct, you will be able to check your work with the answer key. You need to answer at least 26 questions, or 74 percent correct, on your actual FCC exam. Thus, you will want to score 80 percent or greater on your practice exams with Ham Radio Prep so that you feel comfortable going into your test day to attain at least a 74 percent passing grade!

Benefits of the General license

The advantage of upgrading from the Technician to General class is not only additional bands that you can operate on, but also an increase in power. While Techs are limited to 200 watts on HF, Generals can transmit with up to 1,500 watts PEP on the HF bands. That's a lot of power!

The biggest incentive for upgrading from Tech to General is the increased HF privileges. Generals can transmit on two new longwave bands that are below the AM broadcast band, as well as 10 HF bands that allow worldwide communications in a variety of modes. Many hams become a General class licensee because they want those HF privileges.

If you are interested in providing emergency communications, having access to most HF bands is a necessity so that you can pass or receive traffic depending on the propagation at various times on each band. General class licensees have access to all amateur bands, just not every amateur HF frequency. During disaster situations, amateur HF communications get through without failure.

Also, Winlink is an HF-based email system that can be used nearly anywhere in the world. Winlink is just another good reason to get your General license if you are interested in being able to provide a service through your ham station or you have a need to send messages from remote places.

Before you get started

Focus on the **bold** words

These are the correct answers directly from the FCC question pools.

> Happiness lies in the joy of achievement and the thrill of creative effort.

— Franklin Roosevelt

Chapter 1
Commission's Rules

Lesson 1: Commission's Rules

5 Exam questions from this section

The FCC

Remember, the Federal Communications Commission (FCC) is the main authority in charge of amateur radio.

The FCC determines "good engineering and good amateur practice" as applied to the operation of an amateur station in all respects not covered by the Part 97 rules.

Licensing and administering exams

If you succeed in earning your General license, you may administer **Technician only** exams as a volunteer examiner -- if you have a **VEC accreditation**. To become a volunteer examiner, you must be accredited by **a Volunteer Examiner Coordinator** (VEC) and must be at least **18 years** old. For a non-U.S. citizen to become an accredited Volunteer Examiner, **the person must hold an FCC-granted Amateur Radio license of General class or above.**

In order for a Technician class exam to be administered, **at least three General class or higher VEs must observe the examination.**

If you are a Technician class operator and have a Certificate of Successful Completion of Examination (CSCE) that is **valid for 365 days** for General class privileges, you may operate **on any General or Technician class band segment**. You must, however, add the special identifier "AG" after your call sign if the FCC has not yet posted your upgrade on its website **whenever you operate using General class frequency privileges.**

10 years after you pass your General exam, you have a two-year grace period where you can renew your license online without retaking the exam. After the two-year grace period, the license will expire, and **the applicant must pass the current Element 2 exam** before they can receive a new license.

Test help: *Element 2 is the Technician exam.*

Any person who can demonstrate that they once held an FCC-issued General, Advanced or Amateur Extra class license that was not revoked by the FCC may receive partial credit for the elements represented by an expired amateur radio license.

Antennas

The maximum height to which an antenna structure may be erected without requiring notification to the Federal Aviation Administration (FAA) and registration with the FCC is **200 feet,** provided it is not at or near a public use airport.

State and local governments are permitted to regulate amateur radio antenna structures, but they have to **reasonably accommodate Amateur Service communications, and regulations must constitute the minimum practical** to accommodate a legitimate purpose of the state or local entity.

What can you talk about or transmit?

Generally speaking, you can talk about amateur radio, "remarks of a personal character" such as small talk, or emergency/disaster relief.

Selling products on amateur radio generally is prohibited, with a small exception. An amateur station may transmit communications in which the licensee or control operator has a pecuniary (monetary) interest **if other amateurs are being notified of the sale of apparatus normally used in an amateur station and such activity is not done on a regular basis.**

Test help: Basically, what the rules state is that if you want to tell other hams during a net or conversation that you are selling an unwanted antenna that you used in your ham station, it is perfectly legal.

With the exception of emergencies, broadcasting or providing communications to broadcasters also is generally prohibited. An amateur station may provide communications to broadcasters for dissemination to the public only for **the immediate safety of human life or protection of property and there must be no other means of**

communication reasonably available before or at the time of the event.

Generally, transmitting music or secret codes is prohibited. However, music may be transmitted, as well as secret codes when **it is an incidental part of a manned spacecraft retransmission.**

You may use abbreviations or procedural signals in the Amateur Service **if they do not obscure the meaning of a message.**

A one-way transmission that is permitted would be any **transmissions necessary to assist learning the International Morse code.**

Occasional retransmissions of weather and propagation forecast information from U.S. government stations is permitted.

The only language that may be used when identifying your station in making a contact using phone emissions is the **English language.**

Fact: *Basically, you can transmit in any language, however, you must identify with your call sign in English, as usual.*

General license frequency privileges

One of the benefits of earning your General license is access to additional frequencies. Below you will find a chart with all the frequencies available to a General class license holder, as well as the modes available to operate on those frequencies.

You should know what frequencies are in each band, and also that phone and image transmission is prohibited in the **30-meter** band:

General License Privileges

Band	Frequencies	Mode
2200 meters	135.7 - 137.8 kHz	CW, RTTY, Data, Phone, Image (1 watt EIRP maximum)
630 meters	472 - 479 kHz	CW, RTTY, Data, Phone, Image (5 watt EIRP maximum)
160 meters	1.8 - 2.0 MHz	CW, RTTY Data, Phone, Image
80 meters	3.525 - 3.600 MHz	CW, RTTY, Data
	3.800 - 4.000 MHz	CW, Phone, Image
60 meters	5 specific channels at 5 MHz	USB, Phone, CW, RTTY, Data
40 meters	7.025 - 7.125 MHz	CW, RTTY, Data
	7.175 - 7.300 MHz	CW, Phone, Image
30 meters	10.100 - 10.150 MHz	CW, RTTY, Data
20 meters	14.025 - 14.150 MHz	CW, RTTY, Data
	14.225 - 14.350 MHz	CW, Phone, Image
15 meters	21.025 - 21.200 MHz	CW, RTTY, Data
	21.275 - 21.450 MHz	CW, Phone, Image
17, 12, 10 meters		All amateur privileges
Above 50 MHz		All amateur privileges

CW = Morse code
RTTY = Radioteletype

60 meters is restricted to communications only on specific channels, rather than frequency ranges.

A General class license holder is granted all amateur frequency privileges in these bands: **160 meters, 60 meters, 30 meters, 17 meters, 12 meters and 10 meters.**

When choosing a transmitting frequency, the following are good amateur practices:

- Ensure that the frequency and mode selected are within your license class privileges.
- Follow generally accepted band plans agreed to by the amateur radio community.
- Monitor the frequency before transmitting.

All of the above!

When General class licensees are not permitted to use the entire voice portion of a particular band, the portion of the voice segment that is generally available to them would **be the upper frequency end.**

3560 kHz is a frequency that is within the General class portion of the 80-meter band.

3900 kHz is a frequency that is within the General class portion of the 75-meter phone band.

7.250 MHz is a frequency in the General class portion of the 40-meter band (in ITU Region 2).

14305 kHz is a frequency that is within the General class portion of the 20-meter phone band.

21300 kHz is a frequency that is within the General class portion of the 15-meter band.

The following frequencies are available to a control operator who holds a General class license:

- 28.020 MHz
- 28.350 MHz
- 28.550 MHz

All of the above!

2.8 kHz is the maximum bandwidth permitted by the FCC for transmitting on USB frequencies on the 60-meter band.

Communicating with foreign stations

You can communicate with foreign stations not governed by the FCC **when the contact is with amateurs in any country except those whose administrations have notified the ITU that they object to such communications.**

Test help: *The ITU is the International Telecommunication Union, a specialized agency of the United Nations responsible for all matters related to information and communication technologies.*

Third parties

It is sometimes allowable for a third party (another person) who is not licensed to state a message over the radio. However, this is explicitly forbidden in some cases. If a **third party's amateur license has been revoked and not reinstated**, it would disqualify them from participating in stating a message with other amateur radio stations.

Only messages relating to amateur radio or remarks of a personal character, or messages relating to emergencies or disaster relief may be transmitted by an amateur station for a third party in another country.

Interference

Amateur stations may be a secondary user on a band **if they do not cause harmful interference to primary users.**

If you are operating on either the 30-meter or 60-meter bands, and a station in the primary service interferes with your contact, you should **move to a clear frequency or stop transmitting.**

The following situations require an amateur radio operator to take specific steps *(keyword here is specific steps)* to avoid harmful interference:

- When operating within one mile of an FCC monitoring station.
- When using a band where the Amateur Service is secondary.
- When a station is transmitting spread-spectrum emissions.

All of the above!

ITU regions

If you're in the United States, you are currently in International Telecommunication Union (ITU) Region 2. The frequency allocations of ITU **Region 2** apply to radio amateurs operating in North America and South America.

If the FCC has issued you a General class license and you are operating in the 7.175 to 7.300 MHz band, you would be in ITU **Region 2.**

Frequency allocations may differ in areas under FCC jurisdiction outside of ITU Region 2.

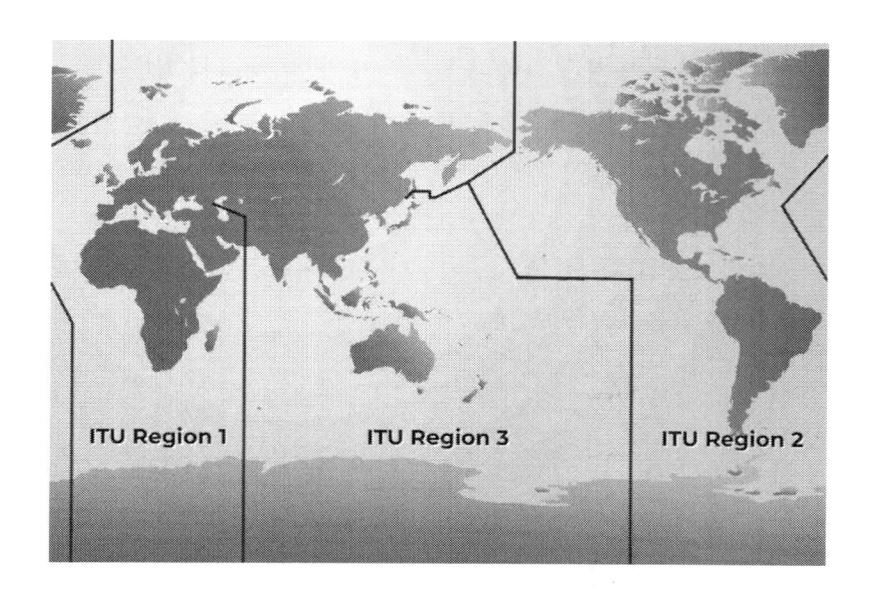

Fact: The Pacific insular areas under FCC jurisdiction in ITU Region 3 are American Samoa, Guam, the Northern Mariana Islands, Baker Island, Howland Island, Jarvis Island, Kingman Reef, Palmyra Island and Wake Island. ITU Region 2 frequencies don't apply to those U.S. islands.

Digital modes and automatically controlled stations

In addition to voice, many ham radio operators use digital modes for communication.

The FCC says that an unattended digital station that transfers messages to and from the internet is called an **automatically controlled digital station.**

The Part 97 limit on the maximum bandwidth occupied by an automatically controlled digital station is **500 Hz.**

If you are operating outside the automatic control band segments, and you wish to conduct communications with a digital station operating under automatic control, your station must be **under local or remote control when initiating contact.**

If you are an engineer or technical expert, it is possible to create your OWN digital protocol! However, you must **publicly document the technical characteristics of the protocol** before using a new digital protocol on the air.

Under no circumstances are messages that are sent via digital modes exempt from Part 97 third-party rules that apply to other modes of communication.

Beacons

The FCC states that the purpose of a beacon station is for **observation of propagation and reception.**

Beacon stations must comply with this condition: **There must be no more than one beacon signal transmitting in the same band from the same station location.**

Test help: *As an example, this is saying that you could have only one beacon station on the air at your station in one band, such as 10 meters. Multiple bands are allowed, however.*

100 watts PEP output is the maximum power limit for beacon stations.

Automatically controlled beacons are permitted on HF frequencies from **28.20 MHz to 28.30 MHz.**

An amateur operator normally should avoid transmitting on 14.100, 18.110, 21.150, 24.930 and 28.200 MHz because **a system of propagation beacon stations operate on those frequencies.**

Repeaters

In the event of interference between a coordinated repeater and an uncoordinated repeater, **the licensee of the uncoordinated repeater has the primary responsibility to resolve the interference.**

The portion of the 10-meter band available for repeater use is

The portion above 29.5 MHz

Repeater use is allowed on the 10-meter band in **the portion above 29.5 MHz**.

Only if a 10-meter repeater control operator holds at least a General class license may a 10-meter repeater retransmit the 2-meter signal from a station that has only a Technician class control operator.

RTTY (radioteletype)

RTTY, or radioteletype, is a digital mode that became popular in amateur radio after World War II, when amateur radio operators inherited teletypes retired from the military. You should learn the following maximum symbol rates for each band when transmitting on RTTY.

Maximum Symbol Rates and Bandwidth		
Band	Symbol Rate (baud)	Bandwidth (KHz)
20-meter or frequencies below 28 MHz	300	1
10-meter	1200	1
6-meter and 2-meter	19.6k	20
1.25-meter and 70 cm	56k	100
30cm and above	No Limit	No Limit

Automatically controlled stations transmitting RTTY, or data emissions, can communicate with other automatically controlled digital stations **anywhere in the 6-meter or shorter wavelength bands, and in limited segments of some of the HF bands.**

Transmitting power rules

When amateur radio operators transmit, they should use **only the minimum power necessary to carry out the desired communications.**

PEP is a measurement specified by FCC rules that regulates maximum power output.

On the exam, they will ask you the maximum transmitting power on different bands. All the correct answers are **1,500 watts PEP output except for** 10.140 MHz, which has a maximum transmitting power of **200 watts PEP output**. So, when in doubt, answer 1,500 watts!

The maximum PEP output allowed for spread-spectrum transmissions is **10 watts.**

The FCC requires that when you are operating in the 60-meter band, and **if you are using an antenna other than a dipole, you must keep a record of the gain of your antenna.**

The maximum power limit on the 60-meter band **is an ERP of 100 watts PEP with respect to a dipole.**

Communicating with Wi-Fi stations

When using modified commercial Wi-Fi equipment to construct an Amateur Radio Emergency Data Network (AREDN), **10 watts** is the maximum allowed PEP transmitter output power.

There is **no part** of the 13-cm band that an amateur station can communicate with non-licensed Wi-Fi stations.

Chapter 1 quiz

1) On which HF/MF bands is a General class license holder granted all amateur frequency privileges?

A. 60 meters, 20 meters, 17 meters and 12 meters
B. 160 meters, 80 meters, 40 meters and 10 meters
C. 160 meters, 60 meters, 30 meters, 17 meters, 12 meters and 10 meters
D. 160 meters, 30 meters, 17 meters, 15 meters, 12 meters and 10 meters

2) Which of the following frequencies is available to a control operator holding a General class license?

A. 28.020 MHz
B. 28.350 MHz
C. 28.550 MHz
D. All of these choices are correct

3) With which of the following conditions must beacon stations comply?

A. A beacon station may not use automatic control
B. The frequency must be coordinated with the National Beacon Organization
C. The frequency must be posted on the internet or published in a national periodical
D. There must be no more than one beacon signal transmitting in the same band from the same station location

4) When General class licensees are not permitted to use the entire voice portion of a band, which portion of the voice segment is generally available to them?

A. The lower frequency end
B. The upper frequency end
C. The lower frequency end on frequencies below 7.3 MHz, and the upper end on frequencies above 14.150
D. The upper frequency end on frequencies below 7.3 MHz, and the

lower end on frequencies above 14.150MHz

5) Which of the following limitations apply to transmitter power on every amateur band?

A. Only the minimum power necessary to carry out the desired communications should be used
B. Power must be limited to 200 watts when using data transmissions
C. Power should be limited as necessary to avoid interference to another radio service on the frequency
D. Effective radiated power cannot exceed 1500 watts

6) When is it permissible to communicate with amateur stations in countries outside the areas administered by the Federal Communications Commission?

A. Only when the foreign country has a formal third-party agreement filed with the FCC
B. When the contact is with amateurs in any country except those whose administrations have notified the ITU that they object to such communications
C. When the contact is with amateurs in any country as long as the communication is conducted in English
D. Only when the foreign country is a member of the International Amateur Radio Union

7) What is the maximum symbol rate permitted for RTTY or data emission transmissions on the 10-meter band?

A. 56 kilobaud
B. 19.6 kilobaud
C. 1200 baud
D. 300 baud

8) On what bands may automatically controlled stations transmitting RTTY or data emissions communicate with other automatically controlled digital stations?

A. On any band segment where digital operation is permitted
B. Anywhere in the non-phone segments of the 10-meter or shorter wavelength bands
C. Only in the non-phone Extra Class segments of the bands
D. Anywhere in the 6-meter or shorter wavelength bands, and in limited segments of some of the HF bands

9) What is the minimum age that one must be to qualify as an accredited Volunteer Examiner?

A. 12 years
B. 18 years
C. 21 years
D. There is no age limit

10) What is the maximum bandwidth permitted by FCC rules for Amateur Radio stations transmitting on USB frequencies in the 60-meter band?

A. 2.8 kHz
B. 5.6 kHz
C. 1.8 kHz
D. 3 kHz

> **Answers to this quiz are located in the answer key at the end of the book.**

> Everything is hard before
> it is easy.

\- Goethe

Chapter 2
Operating Procedures

Lesson 2: Operating Procedures

5 Exam questions from this section

Amateur activities

Amateur volunteers who are formally enlisted to monitor the airwaves for rules violations are part of the **Volunteer Monitoring Program.** The goals and objectives of the Volunteer Monitoring Program are to **encourage amateur radio operators to self-regulate and comply with the rules.** The Volunteer Monitoring Program runs transmitter hunts that help practice the skills for **direction finding used to locate stations violating FCC rules.** If you are interested in joining, you can check the requirements online and apply!

One reason why many amateurs keep a station log is **to help with a reply if the FCC requests information.** Some of the things that are traditionally contained in station logs are:

- Date and time of contact.
- Band and/or frequency of the contact.
- Callsign of station contacted and the signal report given.

All of the above!

Emergency communication

An amateur station is allowed to use any means at its disposal to assist another station in distress **at any time during an actual emergency.**

Except during emergencies, no amateur station has priority access to any frequency.

The first thing you should do if you are communicating with another amateur station and hear a station in distress break in would be to **acknowledge the station in distress and determine what assistance may be needed.**

In emergency situations, the FCC may activate RACES (Radio Amateur Civil Emergency Service) to assist with disaster relief operations. **Only a person holding an FCC-issued amateur operator license** may operate as a RACES control operator. **If the President's War Emergency Powers have been invoked**, the FCC may restrict normal frequency operations of amateur stations participating in RACES.

Whenever sending a distress call in an emergency, you should use **whichever frequency has the best chance of** communicating the distress message.

Calling CQ

Saying "CQ" means "calling any station." If you are looking for a contact with any station, a good way to call CQ on a clear frequency in the HF phone bands is to follow this procedure: **Repeat "CQ" a few times, followed by "this is," then your call sign a few times, then pause to listen, repeating as necessary.**

Example: *"CQ, CQ, this is K1XXX," then pause to listen.*

To answer a station calling CQ when operating phone, state their call sign first followed by your call sign.

Example: *"N1USA, here is K1XXX," then pause for a response from the other station.*

When answering a CQ in Morse code, you should answer **at the fastest speed at which you are comfortable copying, but no faster than the CQ.**

Operations and making contacts

Let's say you hear an interesting conversation when operating on phone and want to jump in the conversation. The recommended way to do this is simply wait for a break between transmissions, and then **say your call sign once.**

VOX operation allows **"hands-free" operation** compared to PTT (push to talk). This is because the radio transmits when it picks up your voice rather than requiring that you push a button.

High levels of atmospheric noise or "static" are typical of the lower HF frequencies during the summer.

To establish contact with a digital messaging system gateway station, you should **transmit a connect message on a station's published frequency.**

When participating in a contest on HF frequencies, you must **identify your station per normal FCC regulations.**

Alpha, Bravo, Charlie, Delta are examples of the NATO Phonetic Alphabet. Here are the rest of the phonetics:

Phonetic Alphabet

Alpha	**J**uliet	**S**ierra
Bravo	**K**ilo	**T**ango
Charlie	**L**ima	**U**niform
Delta	**M**ike	**V**ictor
Echo	**N**ovember	**W**hiskey
Foxtrot	**O**scar	**X**ray
Golf	**P**apa	**Y**ankee
Hotel	**Q**uebec	**Z**ulu
India	**R**omeo	

HAM RADIO PREP

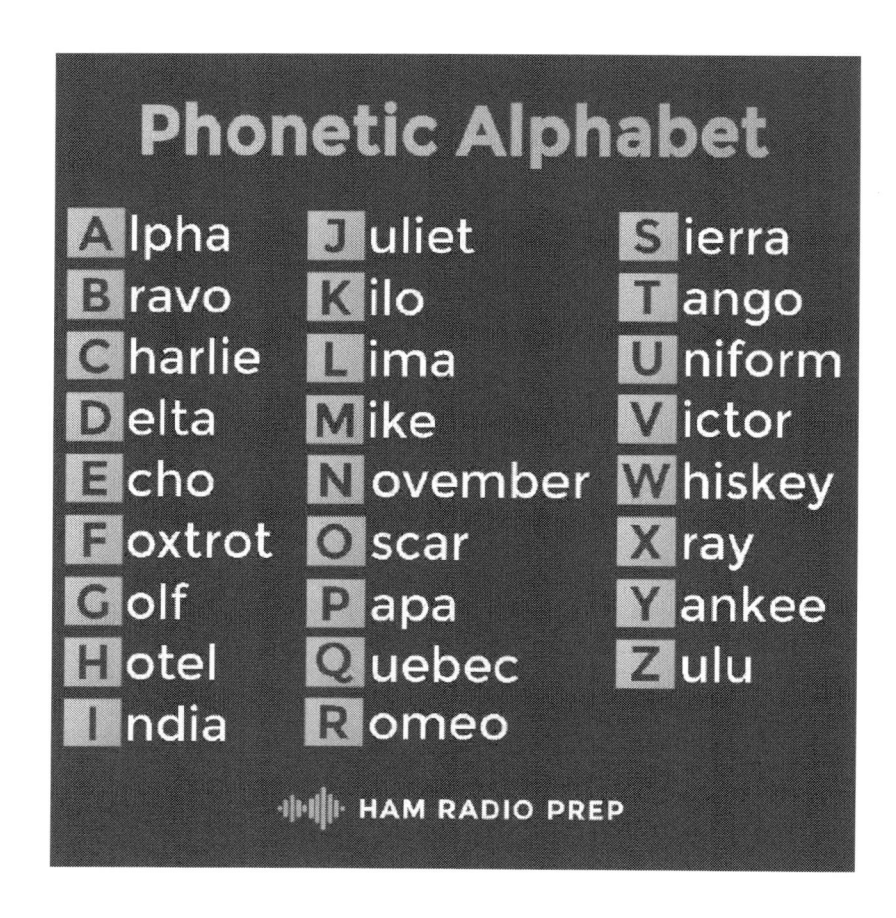

Q signals and codes

Q signals are used to convey specific messages in a clear way. You will need to know the following Q signals and codes for the exam:

Q signal	Meaning
QSL	I acknowledge receipt
QRN	I am troubled by static
QRV	I am ready to receive messages
QRS	Send slower
QRL	"Are you busy?" or "Is this frequency in use?"
AR	Sent to indicate the end of a formal message
QRP	Low power transmit operation
Sending "KN" at the end of a transmission	Listening only for a specific station or stations

Selecting your frequency (band plans)

Whenever you're choosing a frequency to initiate a call, it's best to **follow the voluntary band plan for the operating mode you intend to use.**

A band plan is simply a guide that tells you what mode to use on each frequency to avoid interference. For example, **3570 to 3600 kHz** in the 80-meter band is most commonly used for digital transmission. If you used another mode in this range (such as AM), you likely would cause interference.

The **14.070-14.112 MHz** segment of the 20-meter band is most often used for digital transmissions (avoiding the DX propagation beacons).

When selecting a CW transmitting frequency, the minimum separation should be **150 to 500 Hz** to minimize interference with other stations on adjacent frequencies.

When you match your transmit frequency to the frequency of a received signal in CW operation, that is referred to as **zero beat**.

Talking to foreign stations (DXing)

When a caller is looking for any station outside their own country, they usually use the expression "CQ DX." **Any stations outside the lower 48**

states should respond to a station in the contiguous 48 states that is calling "CQ DX."

There are certain segments in the voluntary band plan reserved for communicating with foreign stations, also called DXing.

For U.S. stations transmitting within the 48 contiguous states, the 50.1 to 50.125 MHz band segment is reserved for **only contacts with stations not within the 48 contiguous states.**

If you want to get into DXing, you may find an azimuthal projection map useful. An azimuthal projection map is **a map that shows true bearings and distances from a particular location.** This can help you set up a directional antenna correctly, especially for long distances.

Antennas

When making contact with another station, the short path is the most direct path to the other station. You also can make contact with the other station via long path by pointing your antenna the opposite direction, **180 degrees from the station's short-path heading.** Surprisingly, long-path contacts sometimes can be much stronger than short path!

FCC rules state that when you are operating within the 60-meter band, **you must keep a record of the gain of your antenna unless you are using a dipole antenna.**

Interference

If propagation changes during your contact and you notice increasing interference from other activity on the same frequency, good amateur practice is to **attempt to resolve the interference problem with the other stations in a mutually acceptable**

manner. In other words, troubleshoot and work it out!

An easy and practical way to avoid harmful interference is by **sending "QRL?" on CW followed by your call sign; or, if using phone, ask whether the frequency is in use, followed by your call sign**.

RST reports

RST reports are a way to communicate to other operators about the quality of their signal (Readability - Strength - Tone).

If you're sending CW (Morse code) and you add a "C" to the RST report, that means a **chirpy or unstable signal**.

Fact: RST reports are given in numbers. For R, or readability, a scale of 1 to 5 is used with 1 meaning unreadable and 5 meaning perfectly readable. Strength (S) and tone (T) are given on a scale of 1 to 9, with 1 being the worst and 9 being the best. T, or tone, is only included in RST reports for CW; phone reports include only readability (R) and strength (S). An RST report for a CW signal might be 599, meaning perfect copy. On phone, a perfect RST report might be given as "59" or "5 and 9" or even "5 by 9."

CW

Full break-in telegraphy (QSK) means that **transmitting stations can receive between code characters and elements.**

If a CW station sends "QRS," you should **send slower.**

When a CW operator sends "KN" at the end of a transmission, it means they are **listening only for a specific station or stations.**

The **AR** prosign is sent to indicate the end of a formal message when using CW.

SSB (single sideband)

Most ham radio operators start by using the FM mode on simplex channels and repeaters on VHF and UHF. SSB, which is really just a special type of AM, is another operating mode that may be a bit more challenging but offers a lot of advantages to the more experienced operator.

One advantage is **less bandwidth used and greater power efficiency** as compared to other analog voice modes on the HF amateur bands. Because of this, **single sideband** is the most commonly used mode on the HF amateur bands.

The transmit audio or microphone gain control is typically adjusted for the proper ALC (automatic level control) setting on an amateur single sideband transceiver.

The customary minimum frequency separation between SSB signals under normal conditions **is approximately 3 kHz.** SSB uses less bandwidth because it only modulates the amplitude *on one side* -- either upper sideband (USB) or lower sideband (LSB). USB and LSB are essentially the same, you just need to learn when to use each.

A true statement of the single sideband voice mode is that **only one sideband is transmitted; the other sideband and carrier are suppressed.**

Upper sideband

Upper sideband is most commonly used for voice communications on the 17-meter and 12-meter bands, and on frequencies 14 MHz and higher.

Upper sideband is most commonly used for SSB voice communications on the VHF and UHF bands.

USB (upper sideband) is the standard sideband used to generate a JT65, JT9 or FT8 digital signal when using AFSK in any amateur band.

Test help: *AFSK is "audio frequency-shift keying," a modulation technique in which digital data is represented by changes in the frequency, or pitch, of an audio tone, and which yields an encoded signal that is suitable for radio transmission.*

Lower sideband

Lower sideband is most commonly used for voice communications on the 160-meter, 75-meter and 40-meter bands because **it is good amateur practice**.

LSB mode is normally used when sending RTTY signals via AFSK with an SSB transmitter (*more on RTTY in the next section*).

49

RTTY (radiotelegraphy)

RTTY (radiotelegraphy) was the first digital transmission mode that gained major acceptance. It uses a form of transmission known as FSK (frequency shift keying).

The most common frequency shift for RTTY emissions in the amateur HF bands is **170 Hz.**

If you cannot decode RTTY or other FSK signals, even though it is apparently tuned in properly, the following might be wrong:

- The mark and space frequencies may be reversed.
- You may have selected the wrong baud rate.
- You may be listening on the wrong sideband.

All of the above!

PSK31

PSK31, or "phase shift keying, 31 Baud" is a very efficient digital mode that is popular among stealth antenna operators because it uses very low power.

Most PSK31 operations on the 20-meter band happen **below the RTTY segment, near 14.070 MHz.**

PACTOR and WINMOR

PACTOR is a modulation mode used by amateurs, marine and military to send and receive digital information. The PACTOR protocol is limited to communication between two stations, and you **cannot** add an additional contact. **Joining an existing contact is not possible; PACTOR connections are limited to two stations.**

WINMOR is a radio protocol used for sending emails across amateur radio and used to implement Winlink. **Winlink** is a communication system that sometimes uses the Internet to transfer messages.

Symptoms that may result from other signals interfering with a PACTOR or WINMOR transmission are:

- Frequent retries or timeouts.
- Long pauses in message transmission.
- Failure to establish a connection between stations.

All of the above!

If you put **the modem or controller in a mode which allows monitoring communications without a connection,** a PACTOR modem or controller can be used to determine whether the channel is in use by other PACTOR stations.

A **DE-9** connector would be a good choice for a serial data port.

Amateur radio software (WSJT/FT8)

WSJT is a computer program used for weak-signal radio communication. It is used to implement the FT8 digital mode, among others. When using the FT8 mode of the WSJT-X software package, **typical exchanges are limited to call signs, grid locators and signal reports.**

One requirement when using the FT8 digital mode is that you must have **computer time accurate within approximately 1 second.**

Chapter 2 quiz

1) Which sideband is most commonly used for voice communications on frequencies of 14 MHz or higher?

A. Upper sideband
B. Lower sideband
C. Vestigial sideband
D. Double sideband

2) Which of the following complies with good amateur practice when choosing a frequency on which to initiate a call?

A. Check to see if the channel is assigned to another station
B. Identify your station by transmitting your call sign at least 3 times
C. Follow the voluntary band plan for the operating mode you intend to use
D. All of these choice are correct

3) Which of the following statements is true of the single sideband voice mode?

A. Only one sideband and the carrier are transmitted; the other sideband is suppressed
B. Only one sideband is transmitted; the other sideband and carrier are suppressed
C. SSB is the only voice mode that is authorized on the 20-meter, 15-meter and 10-meter amateur bands
D. SSB is the only voice mode that is authorized on the 160-meter, 75-meter and 40-meter amateur bands

4) What does the Q signal "QRV" mean?

A. You are sending too fast
B. There Is interference on the frequency
C. I am quitting for the day
D. I am ready to receive messages

5) What is the standard sideband used to generate a JT65, JT9 or FT8 digital signal when using AFSK in any amateur band?

A. LSB
B. USB
C. DSB
D. SSB

6) Which mode is most commonly used for voice communications on the 17-meter and 12-meter bands?

A. Upper sideband
B. Lower sideband
C. Vestigial sideband
D. Double sideband

7) What is the first thing you should do if you are communicating with another amateur station and hear a station in distress break in?

A. Continue your communication because you were on the frequency first
B. Acknowledge the station in distress and determine what assistance may be needed
C. Change to a different frequency
D. Immediately cease all transmissions

8) What could be wrong if you cannot decode an RTTY or other FSK signal even though it is apparently tuned in properly?

A. The mark and space frequencies may be reversed
B. You may have selected the wrong baud rate
C. You may be listening on the wrong sideband
D. All these choices are correct

9) How do you join a contact between two stations using the PACTOR protocol?

A. Send broadcast packets containing your call sign while in MONITOR mode
B. Transmit a steady carrier until the PACTOR protocol times out and disconnects
C. Joining an existing contact is not possible; PACTOR connections are limited to two stations
D. Send a NAK response continuously so that the sending station must stand by

10) What does the Q signal "QRL?" mean?

A. "Will you keep the frequency clear?"
B. "Are you operating full break-in?" or "Can you operate full break-in?"
C. "Are you listening only for a specific station?"
D. "Are you busy?" or "Is this frequency in use?"

> **Answers to this quiz are located in the answer key at the end of the book.**

"

The world is moving so fast these days that the man who says it can't be done is generally interrupted by someone doing it.

- Elbert Hubbard

Chapter 3
Radio Wave Propagation

Lesson 3: Radio Wave Characteristics

3 Exam questions from this section

Atmosphere

Now that you've made it this far, you understand that the atmosphere, especially the ionosphere, is important for the propagation of radio signals. But did you know that the atmospheric layers change between night and day?

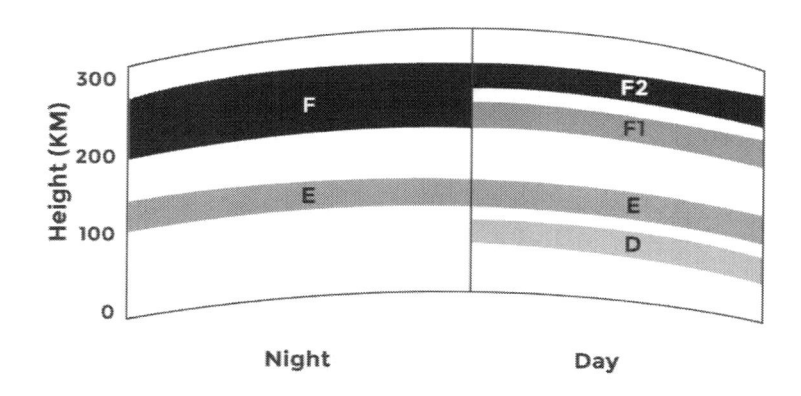

Layers of the ionosphere during the day and night.

At nighttime, the D layer effectively disappears after sunset as electron levels fall quickly and solar radiation is blocked by the Earth. Above the D layer is the E layer, and while ionization levels drop quickly after sunset, a small amount of residual ionization persists at night. Thus, while the E layer virtually disappears at night, there still is residual ionization in the lower parts of the E region that causes some attenuation of signals in the lower portions of the HF spectrum. Also, the F1 and F2 layers combine into a single F layer at night.

The D layer **is the most absorbent of long skip signals during daylight hours on frequencies below 10 MHz. The D layer** also is the ionospheric layer that is closest to the surface of the Earth. Long-distance communications on the 40-meter, 60-meter, 80-meter and 160-meter bands is more difficult during the day because **the D layer absorbs signals at these frequencies during daylight hours.**

Because the E layer is closer to the Earth's surface, the distance that is covered in one hop is a bit shorter. Generally, the approximate maximum distance that normally is covered in one hop using the E layer is **1,200 miles.**

The F2 region is **the highest ionospheric region** and has the longest distance radio wave propagation.

The approximate maximum distance along the Earth's surface that normally is covered in one hop using the F2 region is **2,500 miles.**

Ionospheric layers reach their maximum height **where the sun is overhead.**

During the daytime ionospheric propagation of HF radio waves, a Sudden Ionospheric Disturbance **disrupts signals on lower**

frequencies more than those on higher frequencies.

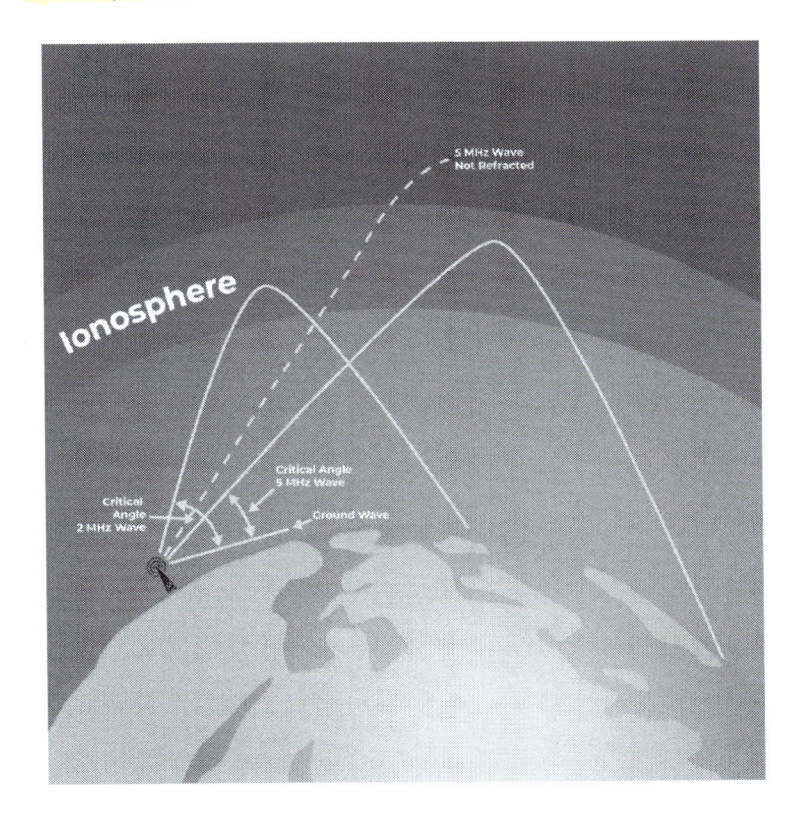

This demonstrates the principle of critical angle and takeoff angles.

Critical angle in radio wave propagation refers to **the highest takeoff angle** that will return a radio wave to Earth under specific ionospheric conditions.

The most effective antenna type for skip communications on 40 meters during the day is **a horizontal dipole placed between 1/8 and 1/4 wavelength above the ground.**

Solar effects

It's always important to consider the effects of the sun on radio wave propagation. One key measurement to consider is the Solar Flux Index (SFI). The Solar Flux Index is a **measure of solar radiation at 10.7 centimeters wavelength.** Generally, a higher SFI is better for HF propagation.

If a solar flare occurs, it can reach Earth in just **8 minutes.** A solar flare causes increased ultraviolet and X-ray radiation with a wide range of effects. Solar flares usually occur near sunspots. The typical sunspot cycle is usually **11 years long**. **Long-distance communication in the upper HF and lower VHF range is**

enhanced when there is a high sunspot number.

The significance of sunspot numbers with regard to HF propagation is that **higher sunspot numbers generally indicate a greater probability of good propagation at higher frequencies.**

In addition to solar flares, the sun also can produce coronal mass ejections that release charged particles. They move slower than increased radiation from a solar flare -- it generally takes about **20 to 40 hours** to affect radio propagation on Earth. When charged particles reach Earth from solar coronal holes, **HF communications are disturbed.**

In addition to these solar events, the regular cycles of **the sun's rotation on its axis** cause HF propagation conditions to vary periodically in a 28-day cycle.

At any point in the solar cycle, the 20-meter band usually can support worldwide propagation during daylight hours.

15 meters, 12 meters and 10 meters are the least reliable for long-distance communications during periods of low solar activity.

Geomagnetic effects

Geomagnetic effects either can help or harm radio communications. A major benefit to radio communications resulting from periods of high geomagnetic activity are **auroras that can reflect** VHF signals.

A geomagnetic storm **is a temporary disturbance in the Earth's magnetosphere**. These geomagnetic storms can cause **degraded high-latitude HF propagation**.

The A-index indicates **the long-term stability of the Earth's geomagnetic field** and the K-index indicates **the short-term stability of the Earth's magnetic field.**

LUF / MUF

LUF stands for the **Lowest Usable Frequency for communications between two points**. MUF stands for the **Maximum Usable Frequency for communications between two points.**

Radio waves with frequencies below the LUF **are completely absorbed by the ionosphere.**

When radio waves with frequencies below the MUF and above the LUF are sent into the ionosphere, **they are bent back to Earth**.

When the LUF exceeds the MUF, **no HF radio frequency will support ordinary** skywave communications over the path.

Factors that affect the MUF:

- Path distance and location.
- Time of day and season.
- Solar radiation and ionospheric disturbances.

All of the above!

On frequencies between 14 and 30 MHz, a reliable way to determine whether the MUF is high enough to support skip propagation between your station and a distant location is to **listen for signals from an international beacon in the frequency range you plan to use.**

When selecting a frequency for lowest attenuation while transmitting on HF, you should **select a frequency just below the MUF.** If signals **are heard on a frequency above the Maximum Usable Frequency (MUF)** on the HF bands, it is a good indication that they are being received via scatter propagation.

Scatter

Scatter is a type of propagation that allows signals to be heard in the transmitting station's skip zone. HF scatter signals in the skip zone usually are weak because **only a small part of the signal energy is scattered into the skip zone.**

A characteristic of HF scatter is **the signals have a fluttering sound.** HF scatter signals often sound distorted because **energy is scattered into the skip zone through several different radio wave paths.**

Near Vertical Incidence Skywave (NVIS)

If you need to transmit over a short distance, but there is some obstacle that prevents a ground wave such as a mountain, you can use a technique called NVIS to reflect signals for short-distance communication. Near Vertical Incidence Skywave (NVIS) propagation is **short distance MF or HF propagation using high elevation angles.**

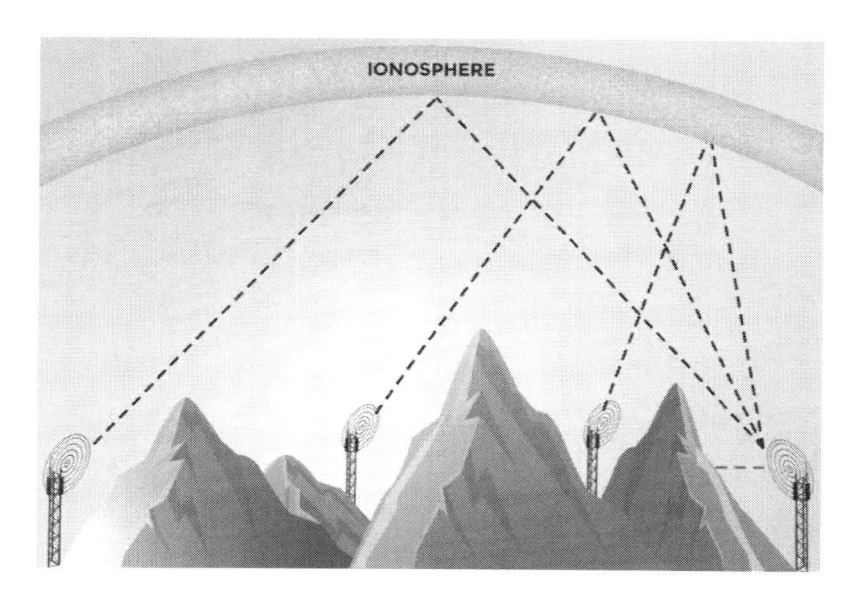

A slightly delayed echo may be heard if skywave signals arrive at your location by both short-path and long-path propagation.

Chapter 3 quiz

1) What effect does a Sudden Ionospheric Disturbance have on the daytime ionospheric propagation of HF radio waves?

A. It enhances propagation on all HF frequencies
B. It disrupts signals on lower frequencies more than those on higher frequencies
C. It disrupts communications via satellite more than direct communications
D. None, because only areas on the night side of the Earth are affected

2) What is the significance of the sunspot number with regard to HF propagation?

A. Higher sunspot numbers generally indicate a greater probability of good propagation at higher frequencies
B. Lower sunspot numbers generally indicate greater probability of sporadic E propagation
C. A zero sunspot number indicates that radio propagation is not possible on any band
D. A zero sunspot number indicates undisturbed conditions

3) What usually happens to radio waves with frequencies below the LUF?

A. They are bent back to Earth
B. They pass through the ionosphere
C. They are completely absorbed by the ionosphere

D. They are bent and trapped in the ionosphere to circle Earth

4) Why are HF scatter signals in the skip zone usually weak?

A. Only a small part of the signal energy is scattered into the skip zone
B. Signals are scattered from the magnetosphere, which is not a good reflector
C. Propagation is through ground waves, which absorb most of the signal energy
D. Propagation is through ducts in the F region, which absorb most of the energy

5) What is a characteristic of skywave signals arriving at your location by both short-path and long-path propagation?

A. Periodic fading approximately every 10 seconds
B. Signal strength increased by 3 dB
C. The signal might be cancelled causing severe attenuation

D. A slightly delayed echo might be heard

6) What benefit can high geomagnetic activity have on radio communications?

A. Auroras that can reflect VHF signals
B. Higher signal strength for HF signals passing through the polar regions
C. Improved HF long path propagation
D. Reduced long delayed echoes

7) What does the A-index indicate?

A. The relative position of sunspots on the surface of the sun
B. The amount of polarization of the sun's electric field
C. The long-term stability of Earth's geomagnetic field
D. The solar radio flux at Boulder, Colorado

8) What is the approximate maximum distance along the Earth's surface that is normally covered in one hop using the E region?

A. 180 miles
B. 1,200 miles
C. 2,500 miles
D. 12,000 miles

9) What is a characteristic of HF scatter?

A. Phone signals have high intelligibility
B. Signals have a fluttering sound
C. There are very large, sudden swings in signal strength
D. Scatter propagation occurs only at night

10) Where on Earth do ionospheric layers reach their maximum height?

A. Where the sun is overhead
B. Where the sun is on the opposite side of Earth
C. Where the sun is rising
D. Where the sun has just set

Answers to this quiz are located in the answer key at the end of the book.

> In radio, you have two tools. Sound and silence.

- Ira Glass

Chapter 4
Amateur Radio Practices

Lesson 4: Amateur Radio Practices

5 Exam questions from this section

Amplifiers

Excessive drive power can lead to permanent damage to a solid-state RF power amplifier.

To reduce distortion due to excessive drive, you should use automatic level control (ALC) with an RF power amplifier.

The correct adjustment for the load or coupling control of a vacuum tube RF power amplifier is **maximum power output without exceeding maximum allowable plate current.**

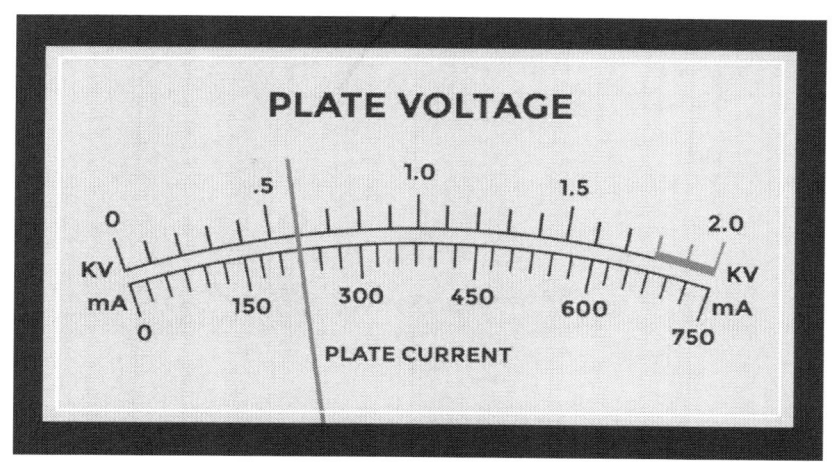

Watch the plate voltage meter for a pronounced dip when adjusting an amplifier.

A pronounced dip on the plate current meter of a vacuum tube RF power amplifier indicates correct adjustment of the plate tuning control.

Antennas

An antenna coupler, or antenna tuner, is a type of device that often is used to match transmitter output impedance to an impedance not equal to 50 ohms.

One disadvantage of using a shortened mobile antenna, as opposed to a full-size antenna, is that the **operating bandwidth may be very limited. To electrically lengthen a physically short antenna,** a capacitance hat is used on a mobile antenna.

The efficiency of electrically short antennas is the biggest limit on an HF mobile installation.

The corona ball sits atop a mobile HF antenna to reduce RF voltage discharge.

To **reduce RF voltage discharge from the tip of the antenna while transmitting,** a corona ball is used on an HF mobile antenna.

Electronic keyer

The purpose of an electronic keyer is **to automate the generation of strings of dots and dashes for CW operation.**

An electronic keyer allows CW to be sent automatically.

Sometimes there is a time delay in a transmitter keying circuit **to allow time for transmit-receive changeover operations to complete properly before RF output is allowed.**

Grounding

If you receive an RF burn when touching your equipment while transmitting on an HF band, assuming the equipment is connected to a ground rod, you might have a problem in that your **ground wire has a high impedance on that frequency.**

Ground loops can be avoided if you **connect all ground conductors to a single point**. One good way to avoid the unwanted effects of stray RF energy in an amateur station is **to connect all equipment grounds together**.

If **you receive reports of a "hum" on your station's transmitted signal,** it could be a symptom of a ground loop somewhere in your station.

High RF voltages on the enclosures of station equipment can be caused by a resonant ground connection.

The metal enclosure of every item of station equipment should be grounded because **it ensures that hazardous voltages cannot appear on the chassis.**

Soldered joints should not be used with the wires that connect the base of a tower to a system of ground rods because **a soldered joint will likely be destroyed by the heat of a lightning strike.**

Measuring equipment

The **antenna and feed line** must be connected to an antenna analyzer when it is being used for SWR measurements.

An antenna analyzer can perform various tasks such as determining coax impedance and SWR.

A common use for an antenna analyzer, other than measuring the SWR of an antenna system, is **determining the impedance of coaxial cable.**

Strong signals from nearby transmitters can affect the accuracy

of measurements when measuring an antenna system with an antenna analyzer.

A field strength meter may be used to monitor relative RF output when making antenna and transmitter adjustments as well as **the radiation pattern of an antenna.**

A standing wave ratio can be determined with a directional wattmeter.

A high input impedance is desirable for a voltmeter because it **decreases the loading on circuits being measured**. An advantage of a digital voltmeter as compared to an analog voltmeter is that it has **better precision for most uses.** However, analog tools are better suited in some cases.

For example, an instance in which the use of an instrument with analog readout may be preferred over an instrument with digital readout would be **when adjusting tuned circuits.**

A two-tone test analyzes the **linearity** of transmitter performance. **Two non-harmonically related audio signals** are used to conduct a two-tone test.

An oscilloscope has multiple uses in an amateur station.

An oscilloscope is the best instrument to use when checking the keying waveform of a CW transmitter. **An oscilloscope** also contains horizontal and vertical channel amplifiers. An advantage of an oscilloscope versus a digital voltmeter is that **complex waveforms can be measured**.

The attenuated RF output of the transmitter is connected to the vertical input of an oscilloscope when checking the RF envelope pattern of a transmitted signal.

S meters

An S meter measures **received signal strength** and is found **in a receiver**.

An S meter gives indication of signal strength on a receiver.

Assuming a properly calibrated S meter, a signal that reads 20 dB over S9 **is 100 times more powerful** than compared to one that reads S9 on a receiver. The power output of a transmitter must be raised **approximately 4 times** to change the S meter reading on a distant receiver from S8 to S9.

Power

The process by which sunlight is changed directly into electricity is called **photovoltaic conversion.** The approximate open-circuit voltage from a fully illuminated silicon photovoltaic cell is **0.5 VDC** (volts DC).

A disadvantage of using wind as the primary source of power for an emergency station is **a large energy storage system is needed to supply power when the wind is not blowing**.

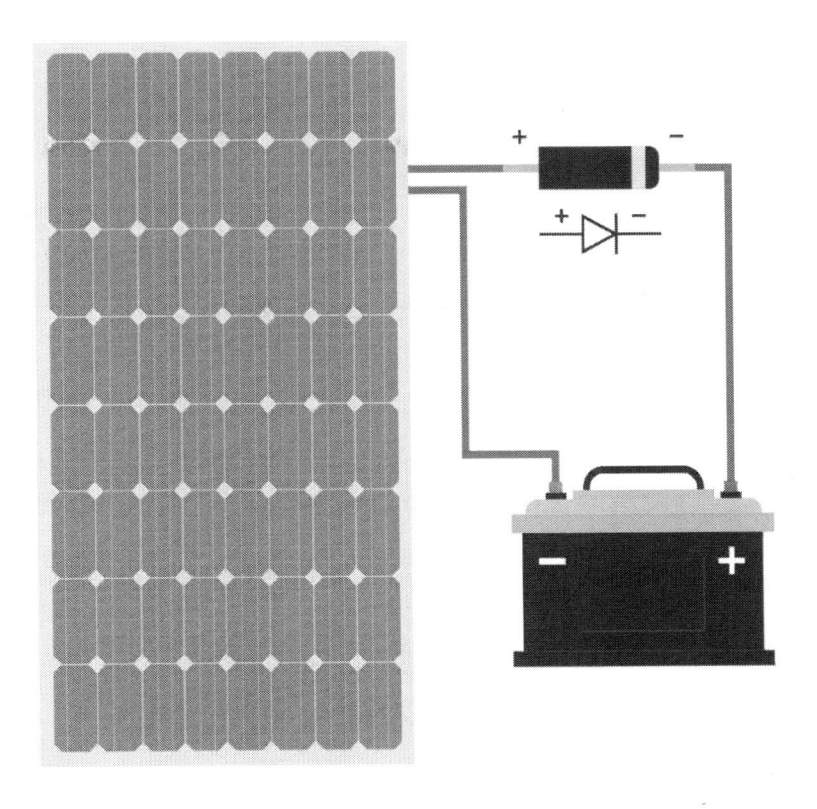

The diode shown at the upper right of this diagram keeps the storage battery from discharging power back through the solar panel.

The reason that a series diode is connected between a solar panel and a storage battery that is being charged by the panel is **the diode prevents self-discharge of the battery through the** panel during times of low or no illumination.

The best direct, fused power connection for a 100-watt HF transceiver from a vehicle's auxiliary power socket is **to the battery using heavy-gauge wire.**

It is best NOT to draw the DC power for a 100-watt HF transceiver from a vehicle's auxiliary power socket because **the socket's wiring may be inadequate for the current drawn by the transceiver.**

Operating

A common use for the dual VFO feature on a transceiver is **to permit monitoring of two different frequencies.**

A transceiver that is in "split" mode is **set to different transmit and receive frequencies.**

Improper action of ALC distorts the signal and can cause spurious emissions if a transceiver's ALC (automatic level control) system is not set properly when transmitting AFSK (audio frequency shift keying) signals with the radio using single-sideband mode.

Symptoms of transmitted RF being picked up by an audio cable carrying AFSK data signals between a computer and a transceiver are:

- The VOX circuit does not unkey the transmitter.
- The transmitter signal is distorted.
- Frequent connection timeouts.

All of the above!

Reducing interference

The purpose of the "notch filter" found on many HF transceivers is **to reduce** interference from carriers in the receiver passband.

A **bypass capacitor** might be useful in reducing RF interference to audio frequency devices.

Distorted speech is heard from an audio device or telephone if there is interference from a nearby single-sideband phone transmitter.

A speech processor affects a transmitted single-sideband phone signal **by increasing the average power.**

One reason to use the attenuator function that is present on many HF transceivers is **to reduce signal overload due to strong incoming signals.**

If there is interference from a nearby CW transmitter on an audio device or telephone system, you will hear **on-and-off humming or clicking.**

It may be possible to reduce or eliminate interference from other signals by selecting the opposite or "reverse" sideband when receiving CW signals on a typical HF transceiver.

To avoid interference from stations very close to the receive frequency, you should use the IF shift control on a receiver.

To reduce RF interference caused by the common-mode current on an audio cable, you could **place a ferrite choke around the cable.**

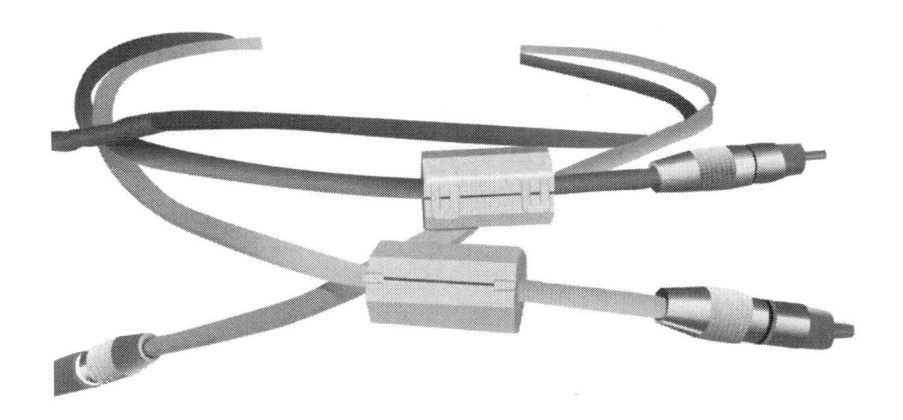

Eliminate interference coming through audio cables by installing ferrite chokes.

Arcing at a poor electrical connection could be a cause of interference covering a wide range of frequencies.

Electrical arcing can come from a variety of sources and cause interference to an amateur radio station.

The following may cause interference to be heard in the receiver of a radio installed in a vehicle:

- The battery charging system.
- The fuel delivery system.
- The vehicle control computer.

All of the above!

A function of a digital signal processor (DSP) is **to remove noise from received signals.** An advantage of a receiver DSP IF filter as compared to an analog filter is that **a wide range of filter bandwidths and shapes can be created.**

A speech processor used in a modern transceiver **increases the intelligibility of transmitted phone signals during poor conditions.**

An incorrectly adjusted speech processor can result in:

- Distorted speech.

- Splatter.
- Excessive background pickup.

All of the above!

A noise blanker works by **reducing receiver gain during a noise pulse.**

Some of the controls on the front of an HF amateur transceiver include NB for noise blanker, which reduces receiver gain, and ATT for attenuator, which helps reduce signal overload from strong signals being received.

Received signals may become distorted when the noise reduction control level in a receiver is increased.

Bonding all equipment enclosures together is a technique that helps to minimize RF "hot spots" in an amateur station.

SSB (single sideband)

When using upper sideband (USB), you should set the carrier frequency at the low end of the frequency. When using lower sideband (LSB), it is the opposite -- you should set your carrier frequency to the higher edge.

When using 3-kHz wide LSB, your displayed carrier frequency should be **at least 3 kHz above the edge of the segment.**

When using 3-kHz wide USB, your displayed carrier frequency should be **at least 3 kHz below the edge of the band.**

The frequency range occupied by a 3-kHz LSB signal when the displayed carrier frequency is set to 7.178 MHz **is 7.175 to 7.178 MHz.**

The frequency range is occupied by a 3-kHz USB signal with the displayed carrier frequency set to 14.347 MHz **is 14.347 to 14.350 MHz.**

Test help: *Just subtract when it's lower sideband (LSB) and add when it's upper sideband (USB).*

Chapter 4 quiz

1) What is one advantage of selecting the opposite, or "reverse," sideband when receiving CW signals on a typical HF transceiver?

A. Interference from impulse noise will be eliminated
B. More stations can be accommodated within a given signal passband
C. It may be possible to reduce or eliminate interference from other signals
D. Accidental out-of-band operation can be prevented

2) Which of the following is a use for the IF shift control on a receiver?

A. To avoid interference from stations very close to the receive frequency
B. To change frequency rapidly
C. To permit listening on a different frequency from that on which you are transmitting

D. To tune in stations that are slightly off frequency without changing your transmit frequency

3) What item of test equipment contains horizontal and vertical channel amplifiers?

A. An ohmmeter
B. A signal generator
C. An ammeter
D. An oscilloscope

4) What is an advantage of a digital voltmeter as compared to an analog voltmeter?

A. Better for measuring computer circuits
B. Better for RF measurements
C. Better precision for most uses
D. Faster response

5) What signals are used to conduct a two-tone test?

A. Two audio signals of the same frequency shifted 90 degrees

B. Two non-harmonically related audio signals
C. Two swept frequency tones
D. Two audio frequency range square wave signals of equal amplitude

6) Which of the following instruments may be used to monitor relative RF output when making antenna and transmitter adjustments?

A. A field strength meter
B. An antenna noise bridge
C. A multimeter
D. A Q meter

7) Which of the following would reduce RF interference caused by common-mode current on an audio cable?

A. Placing a ferrite choke around the cable
B. Adding series capacitors to the conductors
C. Adding shunt inductors to the conductors

D. Adding an additional insulating jacket to the cable

8) Which of the following can be the result of an incorrectly adjusted speech processor?

A. Distorted speech
B. Splatter
C. Excessive background pickup
D. All these choices are correct

9) What could be a symptom of a ground loop somewhere in your station?

A. You receive reports of "hum" on your station's transmitted signal
B. The SWR reading for one or more antennas is suddenly very high
C. An item of station equipment starts to draw excessive amounts of current
D. You receive reports of harmonic interference from your station

10) Which of the following most limits an HF mobile installation?

A. "Picket fencing"
B. The wire gauge of the DC power line to the transceiver
C. Efficiency of the electrically short antenna
D. FCC rules limiting mobile output power on the 75-meter band

Answers to this quiz are located in the answer key at the end of the book.

> Success is no accident. It is hard work, perseverance, learning, studying, sacrifice and most of all, love of what you are doing or learning to do.

— Pele

Chapter 5
Electrical Principles

Lesson 5: Electrical Principles

3 Exam questions from this section

Definitions

Impedance is **the opposition to the flow of current in an AC circuit** and is measured in ohms.

Impedance is caused by resistance in the circuit because of elements such as lights, resistors, etc. However, circuits also have capacitive and inductive elements.

Reactance is **the opposition to the flow of alternating current caused by capacitance or inductance** and ALSO is measured in **ohms**.

Voltage can be symbolized by a V or an E, as you will see in this lesson.

Inductors / capacitors

Reactance causes opposition to the flow of alternating current in a capacitor **AND** an inductor.

An inductor reacts to AC in this way: **As the frequency of the applied AC increases, the reactance increases.**

Test help: *Inductors increase … "I" is for increase!*

Here's how a capacitor reacts to AC: **As the frequency of the applied AC increases, the reactance decreases.**

Units

Here are a few quick notes on farads, the units that measure capacitance:

A nanofarad is by definition 10^{-9} farads. A picofarad is 10^{-12} farads. This means that a picofarad is one-thousandth of a nanofarad. Therefore, 22,000 picofarads are equivalent to **22 nF.**

A microfarad (µF) is equal to 10^{-6} farads. A nanofarad is equal to 10^{-9} farads. Therefore, a microfarad is 1000 times larger than a nanofarad. Thus, 4700nF / 1000 = **4.7 µF.**

A change of **approximately 3 dB** represents a factor of two increase or decrease in power.

Series / parallel

When two elements are connected in a line so that current flows through one path through them, they are in series.

When two resistors are connected in series, the resistances add together!

You can see here that when elements are in series, there is only one path for current to flow through them. This means that the resistances just add together. **A series resistor** is a component that increases the total resistance of a resistor.

If you have three equal resistors and the total resistance is 450 ohms, you can just divide by three to get the value of each resistor:

450 ohms / 3 = **150 ohms**

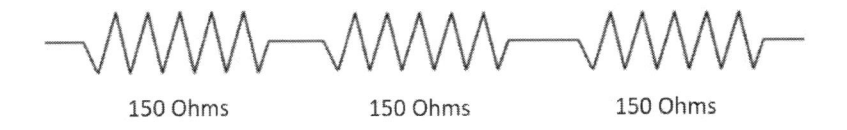

150 Ohms   150 Ohms   150 Ohms

Three equal resistors: 150 + 150 + 150 = 450 ohms

When two elements are connected in parallel, there are two paths for the current to flow.

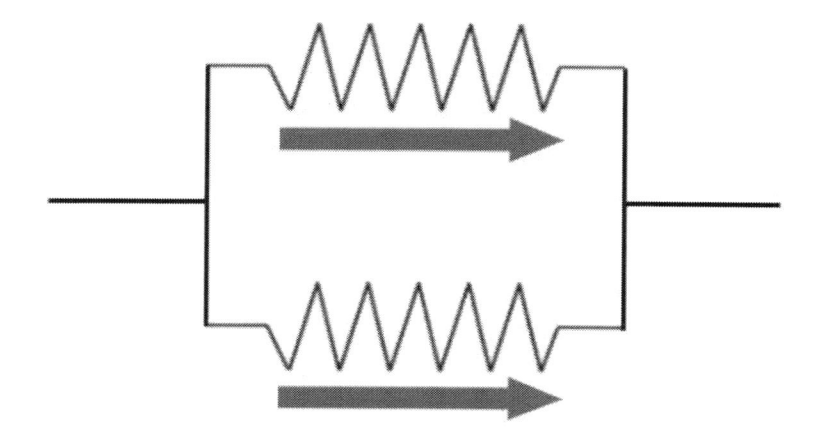

When two resistors are connected in parallel, there are more paths and the resistance gets smaller.

The total current **equals the sum of the currents through each branch** in a purely resistive parallel circuit. This just means we add the current through each of the arrows in the graphic before this paragraph, and it gives us the total current.

When you place resistors in parallel, the resistance decreases. For example, if there are three 100-ohm resistors in parallel, the total resistance is 100 ohms / 3 = **33.3 ohms.** It gets a little more complex when the resistors have different values.

Imagine we have three resistors in parallel, with the values of 10 ohms, 20 ohms and 50 ohms. This case is a bit complex. The total resistance of resistors that are in parallel is the reciprocal of the total resistance equal to the sum of the individual resistors' reciprocal values.

We use the following formula:

$$R = \frac{1}{\frac{1}{R1}+\frac{1}{R2}+\frac{1}{R3}}$$

To solve for the total resistance, we plug in the values of our three resistors:

$$R_{Total} = \cfrac{1}{\cfrac{1}{10\ \Omega} + \cfrac{1}{20\ \Omega} + \cfrac{1}{50\ \Omega}}$$

$$= \cfrac{1}{\cfrac{10}{100\ \Omega} + \cfrac{5}{100\ \Omega} + \cfrac{2}{100\ \Omega}}$$

$$= \cfrac{1}{\cfrac{17}{100\ \Omega}}$$

$$= \cfrac{100}{17}\ \Omega$$

So, our answer = **5.9 ohms**

When you take the test, you also can answer this question by process of elimination. If you are combining three resistors in parallel, the resistance is going to get lower. This gets rid of two wrong answers. The other wrong answer is .17 ohms -- this is way too low! **5.9 ohms is the only answer that is lower resistance, but not so low that it doesn't make sense.**

What happens when we connect other elements in parallel, like inductors? Well, inductors in the series act just like resistors. They add together.

An inductor in series should be added to an inductor to increase the inductance.

An inductor in series

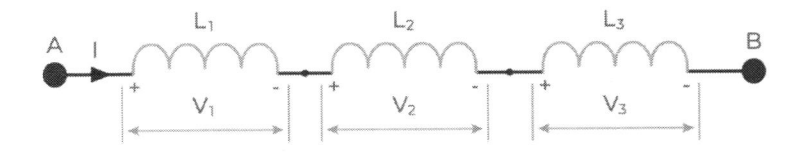

Should be added to increase the inductance

The total inductance of inductors in series is simply the sum of the individual inductances.

L = L1 + L2 + L3 + ... so the inductance of a 20-millihenry inductor connected in series with a 50-millihenry inductor is **70 millihenrys.**

When inductors are connected in parallel, the inductance decreases, just like resistors. Three 10 millihenry inductors connected in parallel would divide the inductance by a factor of 3: 10 mH/3= **3.3 millihenrys.**

Capacitors are exactly the opposite of inductors in series and parallel circuits. Capacitors in parallel add together, increasing the capacitance. In other words, **a capacitor in parallel** should be added to a capacitor to increase the capacitance.

What is the total capacitance of two 5-nanofarad capacitors and one 750-picofarad capacitor in parallel? First you need to convert picofarads to nanofarads so the units match.

750 picofarads = .75 nanofarads. Now we can just add!

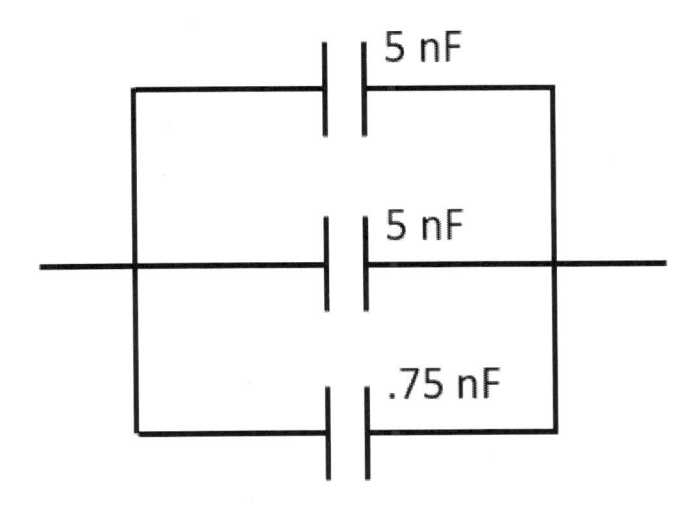

Capacitors in parallel add together, essentially creating a bigger capacitor!

5.0 nanofarads + 5.0 nanofarads + 0.75 nanofarads = **10.750 nanofarads**

When capacitors are in series, the total capacitance *decreases*. When all the capacitors are equal in value, you may calculate the total capacitance by dividing the common capacitor value by the number of capacitors in the series. For example, consider 3 capacitors of 100 microfarads in series.

100 microfarad / 3 = **33.3 microfarads**

To calculate the total capacitance for capacitors in series when the capacitors are not equal, you need to use this specific formula. You will notice it is the same format as calculating the total resistance of parallel resistors! The total capacitance of capacitors in series is the reciprocal of the sum of the individual capacitor's reciprocal values.

That creates this formula for two capacitors:

$$C = \frac{1}{\frac{1}{C1} + \frac{1}{C2}}$$

Let's try connecting a 20-microfarad capacitor in series with a 50-microfarad capacitor, using this formula:

$$C = \frac{1}{\frac{1}{20\ \mu F} + \frac{1}{50\ \mu F}}$$

$$= \frac{1}{\frac{5}{100} + \frac{2}{100}}$$

$$= \frac{1}{\frac{7}{100}}$$

$$= \frac{100}{7}$$

= 14.3 microfarads

When you're taking the exam, you can use the process of elimination for this question, too. You know that when capacitors are in series, the total capacitance will get smaller. That eliminates two of the wrong answers. Starting with 20 microfarads and 50 microfarads, **14.3 microfarads** is the only value that makes sense.

Ohm's Law

Ohm's Law is the fundamental law of circuits, and it relates voltage, current and resistance. We will need this formula:

Voltage (E) equals current (I) multiplied by resistance (R). Voltage can also be written with a V:

V = I x R

As in the Technician course, we can use the triangle to calculate voltage, resistance or current.

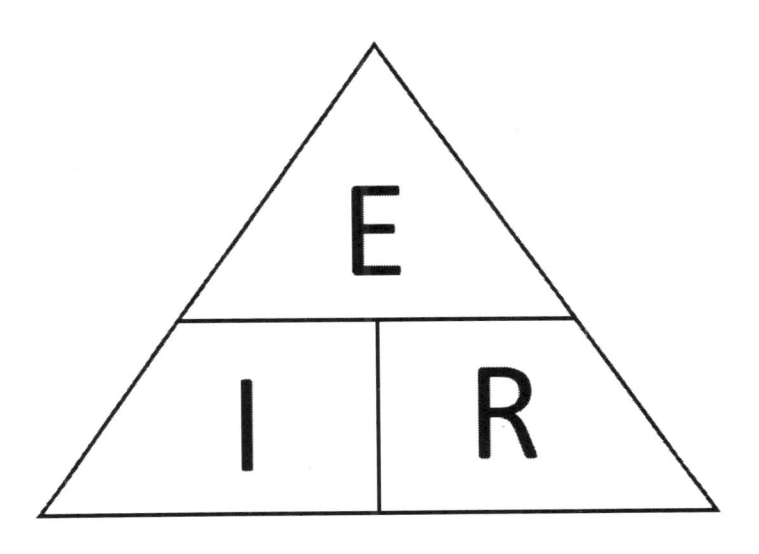

We can circle the "I" to calculate current, and we see that current (I) equals voltage (E) divided by resistance (R):

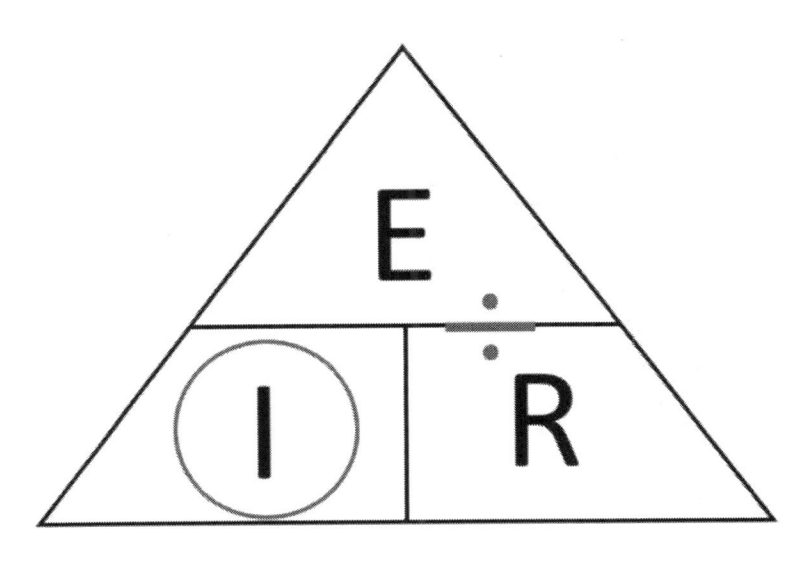

To answer the next questions, we will need to combine this formula with new formulas for power and voltage.

Power

Let's take a look at the power equation now.

Power equals voltage times current and is measured in watts.

P = V x I

Test help: *More voltage and more current means more power!*

If the test gives you voltage and current and asks you to find power, you can plug these numbers directly into the power formula. How many watts of electrical power are used by a 12-VDC light bulb that draws 0.2 amperes?

(12V) x (0.2 amps) = **2.4 watts**

The test expects you to be able to combine this equation with the Ohm's Law equation to solve more complex problems.

The test gives you this question: How many watts of electrical power is used if 400 VDC is supplied to an 800-ohm load?

We need to calculate power, but this time we have voltage and resistance. Let's start with the power equation and plug in the voltage.

P = 400V x I

However, we need the current (I) to finish the equation! We can go back to Ohm's Law and calculate the current (I) using the voltage and resistance.

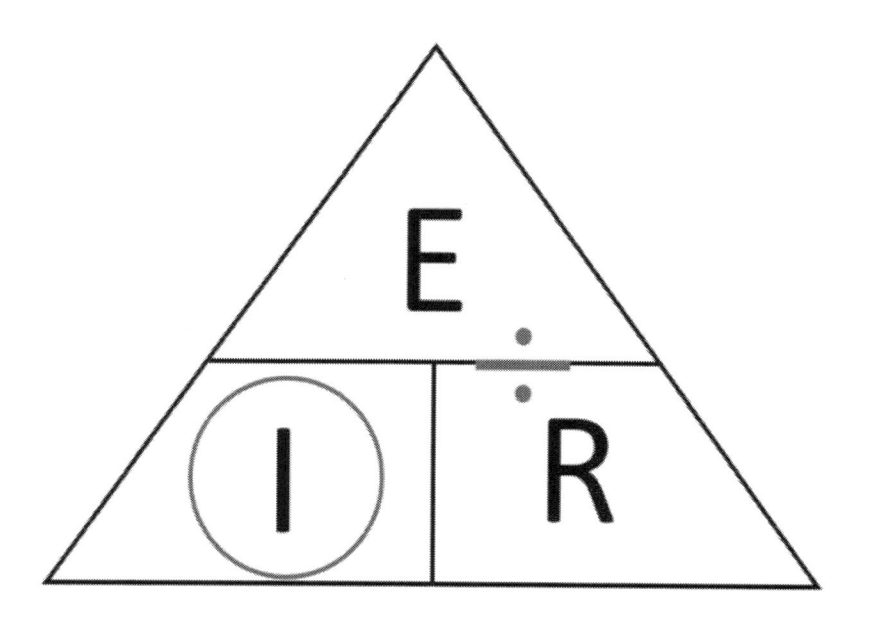

I = E/R

We plug in the values from the problem in this equation:

I = 400V / 800 ohms

I = 0.5 amps

Now back to our power equation, we can insert the current I:

P = 400V x 0.5 amps = **200 watts.**

Here is another question that needs us to combine the formulas: How many watts are dissipated when a current of 7.0 milliamperes flows through 1,250 ohms resistance?

For starters:

P = power in watts
I = current in amperes
E = energy in volts
R = resistance in ohms

First, we're told:

I current = 7.0 milliamperes, which equals 0.007 amperes

R resistance = 1.25 kilohms = 1,250 ohms

So, in order to calculate for power, we use the equation:

P = I x E and then combine it with the Ohm's Law equation E = I x R

This gives you a formula of P = I x (I x R)

Now we plug in the numbers:

P = (0.007 A) × (0.007 A × 1250 Ω) = 0.061 watts, which is 61 milliwatts

or:

P = (0.007A)2 × 1250 Ω = 0.061 watts = 61 milliwatts

Thus, you can see the answer is **approximately 61 milliwatts.**

Peak envelope power (PEP)

Peak envelope power is the highest power that is supplied to an antenna during a cycle -- the highest power the antenna receives.

Unmodulated carriers

If the power supplied is steady ("unmodulated"), then the peak envelope power is the same as the average power. A carrier that always delivers the same power is said to be unmodulated.

What is the ratio of peak envelope power to average power for an unmodulated carrier? It is **1.00** because the values are the same.

If the output PEP of an unmodulated carrier with an average reading wattmeter connected to the transmitter is indicating 1,060 watts, then the output PEP would be the same, **1,060 watts**!

Calculating PEP with peak-to-peak voltage

Calculating PEP using the peak-to-peak voltage, we must use the following formulas:

Peak voltage = Peak-to-peak voltage / 2

RMS = Peak voltage x .707

P = V x I (you must use RMS voltage in this equation, not peak!)

V = I x R (Ohm's Law)

What is the output PEP from a transmitter if an oscilloscope measures 200 volts peak-to-peak across a 50-ohm dummy load connected to the transmitter output?

Peak voltage = 200 / 2 = 100V

RMS Voltage = 100V x .707 = 70.7 V

I = 70.7V / 50 ohms = 1.414 amps

P = 70.7 V x 1.414 amps = **100 watts**

What is the output PEP from a transmitter if an oscilloscope measures 500 volts peak-to-peak across a 50-ohm resistive load connected to the transmitter output? **625 watts**

Calculating RMS voltage with resistance and power

P = V^2 / R

The RMS voltage across a 50-ohm dummy load dissipating 1,200 watts **is 245 volts**.

What we know: P = 1,200W, R = 50 ohms

1,200 = V^2 / 50 ohms

V = sqrt(1,200 x 50) = **245 volts**

20.6 percent is the percentage of power loss resulting from a transmission line loss of 1 dB.

RMS and PEV

To find the peak envelope voltage (PEV), multiply the RMS by the square root of 2 (which is approximately 1.414).

PEV = RMS x 1.414

For example, to find the peak-to-peak voltage of a sine wave that has an

RMS voltage of 120 volts, the formula would look like this:

PEV = 120 x 1.414 = 169.7 V

Now to find the peak-to-peak voltage, you would multiply the PEV by 2 times so:

Peak-to-peak = 169.7 volts x 2 = **339.4 volts**

To find the RMS voltage, multiply the peak voltage by 0.707.

RMS = PEV x .707

RMS voltage = Peak voltage x 0.707

So, for example the RMS voltage of a sine wave with a value of 17 volts peak would be:

17 volts x 0.707 = **12 volts**

Transformers

Transformers, like the ones you see on the utility poles outside your home, can transform voltage by using two inductors side by side. At its core, an inductor is just a coil of wires.

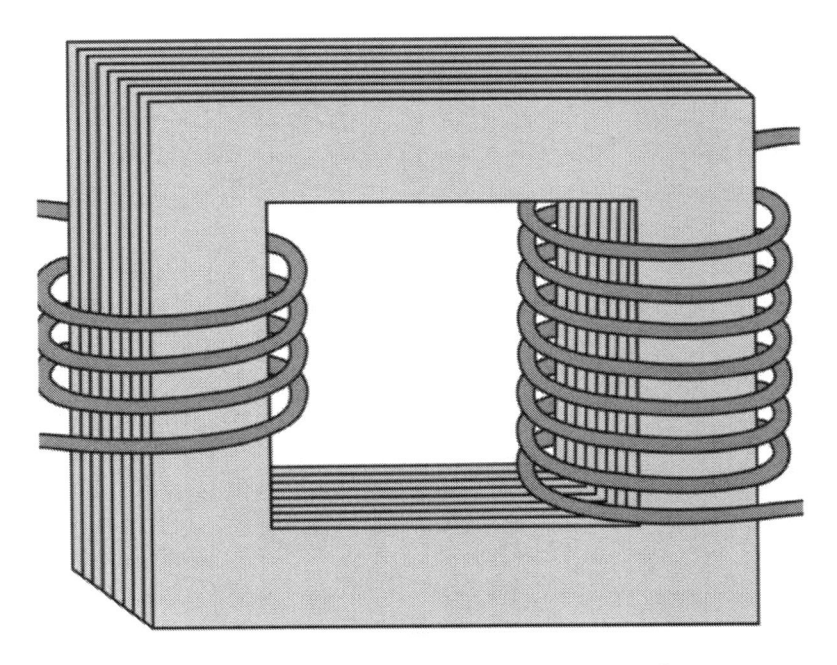

A transformer is made of two paired inductors.

Transformers can transform voltage from one inductor to the other inductor so that the voltage on the other side is bigger or smaller.

Mutual inductance causes a voltage to appear across the secondary winding of a transformer when an AC voltage source is connected across its primary winding.

The conductor of the primary winding of many voltage step-up transformers is larger in diameter than the conductor of the secondary winding **to accommodate the higher current of the primary.**

The output voltage is multiplied by 4 when a signal is applied to the secondary winding of a 4:1 voltage step-down transformer instead of the primary winding.

This seems easy, 4 times is the same as 4:1. But how do we calculate for other transformers? The key is in the number of turns (N). We can use the number of turns to calculate the voltage using this ratio:

$$\frac{N1}{N2} = \frac{V1}{V2}$$

What is the RMS voltage across a 500-turn secondary winding in a transformer, if the 2,250-turn primary is connected to 120 VAC?

N1 = 2,250 turns, N2 = 500 turns, V1 = 120V, **V2 = ?**

$$\frac{2250\ turns}{500\ turns} = \frac{120V}{V2}$$

$$2250 \times V2 = 120 \times 500$$

$$V2 = \frac{60000}{2250} = 26.67V$$

The answer is **26.7 volts**

The turns ratio of a transformer used to match an audio amplifier having 600-ohm output impedance to a speaker having 4-ohm impedance is **12.2 to 1.**

Impedance matching

The reason why impedance matching is important is **so the source can deliver maximum power to the load**. When the impedance of an electrical load is equal to the output impedance of a power source, assuming both impedances are resistive, **the source can deliver maximum power to the load.**

Inserting an LC network between two circuits is one method of impedance matching between two AC circuits.

One reason to use an impedance matching transformer is **to maximize the transfer of power.**

The following devices can be used for impedance matching at radio frequencies:

- A transformer.
- A Pi-network.
- A length of transmission line.

All of the above!

Chapter 5 quiz

1) Which of the following causes opposition to the flow of alternating current in an inductor?

A. Conductance
B. Reluctance
C. Admittance
D. Reactance

2. What unit is used to measure reactance?

A. Farad
B. Ohm
C. Ampere
D. Siemens

3) How does an inductor react to AC?

A. As the frequency of the applied AC increases, the reactance decreases
B. As the amplitude of the applied AC increases, the reactance increases
C. As the amplitude of the applied AC increases, the reactance decreases
D. As the frequency of the applied AC increases, the reactance increases

4) Which of the following describes one method of impedance matching between two AC circuits?

A. Insert an LC network between the two circuits
B. Reduce the power output of the first circuit
C. Increase the power output of the first circuit
D. Insert a circulator between the two circuits

5) What dB change represents a factor of two increase or decrease in power?

A. Approximately 2 dB
B. Approximately 3 dB
C. Approximately 6 dB
D. Approximately 12 dB

6) What value of an AC signal produces the same power dissipation in a resistor as a DC voltage of the same value?

A. The peak-to-peak value
B. The peak value
C. The RMS value
D. The reciprocal of the RMS value

7) What percentage of power loss would result from a transmission line loss of 1 dB?

A. 10.9 percent

B. 12.2 percent
C. 20.6 percent
D. 25.9 percent

8) What is the output PEP from a transmitter if an oscilloscope measures 500 volts peak-to-peak across a 50-ohm resistive load connected to the transmitter output?

A. 8.75 watts
B. 625 watts
C. 2500 watts
D. 5000 watts

9) What is the turns ratio of a transformer used to match an audio amplifier having 600-ohm output impedance to a speaker having 4-ohm impedance?

A. 12.2 to 1
B. 24.4 to 1
C. 150 to 1
D. 300 to 1

10) Why is the conductor of the primary winding of many voltage step-up transformers larger in diameter than the conductor of the secondary winding?

A. To improve the coupling between the primary and secondary
B. To accommodate the higher current of the primary
C. To prevent parasitic oscillations due to resistive losses in the primary
D. To ensure that the volume of the primary winding is equal to the volume of the secondary winding

> **Answers to this quiz are located in the answer key at the end of the book.**

> When everything else fails, amateur radio oftentimes is our last line of defense. ...When you need amateur radio, you really need them.

— Craig Fugate, FEMA administrator

Chapter 6
Circuit Components

Lesson 6: Circuit Components

2 Exam questions from this section

Types of batteries

The minimum allowable discharge voltage for maximum life of a standard 12-volt lead acid battery is **10.5 volts.**

The high discharge current is an advantage of the low internal resistance of nickel-cadmium batteries.

It is **never** acceptable to recharge a carbon-zinc primary cell.

Fact: Zinc carbon primary cells are not designed to be charged. If charged, there is a high risk of leakage or explosion of the battery.

Resistors

Resistors heat up and experience energy loss because of the heat. When the temperature of a resistor is increased, the resistance **will change depending on the resistor's temperature coefficient**.

If you are building an RF circuit, you probably want to avoid wire-wound resistors. Wire-wound resistors should not be used in RF circuits because **the resistor's inductance could make circuit performance unpredictable.**

Capacitors

There are several different types of capacitors, and each has advantages and disadvantages depending on the application.

One advantage of ceramic capacitors is their **comparatively low cost.**

An advantage of electrolytic capacitors is their **high capacitance for a given volume.**

The polarity of applied voltages is important for polarized capacitors for many reasons:

- Incorrect polarity can cause the capacitor to short circuit.
- Reverse voltages can destroy the dielectric layer of an electrolytic capacitor.
- The capacitor could overheat and explode.

All of the above!

Inductors

Like capacitors, there also are several types of inductors.

There are many advantages of using a ferrite core toroidal inductor:

- Large values of inductance may be obtained.
- The magnetic properties of the core may be optimized for a specific range of frequencies.
- Most of the magnetic field is contained in the core.

All of the above!

The biggest determination of the performance of a ferrite core at different frequencies is **the composition, or "mix," of materials used.**

A ferrite bead or core reduces common-mode RF current on the shield of a coaxial cable **by creating an impedance in the current's path.**

The winding axes of two solenoid inductors should be oriented **at right angles to each other** to minimize their mutual inductance.

When an inductor is operated above its self-resonant frequency, **it becomes capacitive.**

One reason not to use wire-wound resistors in an RF circuit is **the resistor's inductance could make circuit performance unpredictable.**

Transistors

Transistors are an extremely important circuit component that are used for amplification, logic and much more.

Triode vacuum tubes were invented before the modern transistor still but are used today. The **control grid** of a triode vacuum tube is used to regulate the flow of electrons between cathode and plate.

The primary use of a screen grid in a vacuum tube is **to reduce grid-to-plate capacitance.**

A **field effect transistor** is a solid-state device most like the vacuum tube in its general operating characteristics.

MOSFETs are a common type of field effect transistor. In the construction of a MOSFET, **the gate is separated from the channel with a thin insulating layer.**

The stable operating points for a bipolar transistor used as a switch in a logic circuit are **its saturation and cutoff regions.**

Diodes

The approximate junction threshold voltage of a germanium diode is **0.3 volts.**

The approximate junction threshold voltage of a conventional silicon diode is **0.7 volts.** Silicon diodes are by far the most common.

Lower capacitance is an advantage of using a Schottky diode rather than a standard silicon diode in an RF switching circuit.

LEDs are a special kind of diode that emit light when current passes through them. That's why they are called light-emitting diodes. LEDs are so common, you probably have them in your house, in your phone and even in your television and computers. They also are used in circuits for ham radio gear.

Using an LED indicator has many advantages compared to an incandescent indicator:

- Lower power consumption.
- Faster response time.
- Longer life.

> *All of the above!*

No wonder they are so common.

When an LED is emitting light, it is **forward biased**. This is because the current is passing forward through it.

A characteristic of a liquid crystal display (LCD) is **it utilizes ambient or back lighting.**

Integrated circuits

Putting various components together on a chip will create an integrated circuit. By connecting all these components together in the correct way, we are able to create amazing devices with circuits.

We can separate them into two kinds -- digital or analog. Digital is logic based -- ones and zeros. Analog is signal based, like analog tapes.

A **linear voltage regulator** is an analog integrated circuit.

An integrated circuit operational amplifier is another **analog** device.

MMIC means **Monolithic Microwave Integrated Circuit.**

CMOS (Complementary Metal Oxide Semiconductor) integrated circuits have **low power consumption** compared to TTL (transistor-transistor logic) integrated circuits.

In non-volatile memory, **the stored information is maintained even if power is removed.**

ROM is **Read Only Memory**.

A microprocessor is **a computer on a single integrated circuit.** They are small, but capable of running programs!

Connectors

If you haven't picked up on it yet, choosing the correct connectors is very important in amateur radio. Even the best components will perform badly if the connectors are not correct.

The **computer and transceiver** of an amateur radio station often are connected using a USB interface.

A **DE-9** connector would be a good choice for a serial data port.

A **PL-259** connector is commonly used for RF connections at frequencies up to 150 MHz.

RCA Phono connectors are commonly used for audio signals in amateur radio stations.

The main reason to use keyed connectors instead of non-keyed types is to **reduce the chance of incorrect mating.**

A type N connector is **a moisture-resistance RF connector useful to 10 GHz.**

A DIN-type connector is **a family of multiple circuit connectors suitable for audio and control signals.**

A type SMA connector is **a small threaded connector suitable for signals up to several GHz.**

Chapter 6 quiz

1) What is the minimum allowable discharge voltage for maximum life of a standard 12-volt lead-acid battery?

A. 6 volts
B. 8.5 volts
C. 10.5 volts
D. 12 volts

2) Which of the following is a reason not to use wire-wound resistors in an RF circuit?

A. The resistor's tolerance value would not be adequate for such a circuit
B. The resistor's inductance could make circuit performance unpredictable
C. The resistor could overheat
D. The resistor's internal capacitance would detune the circuit

3) What are the stable operating points for a bipolar transistor used as a switch in a logic circuit?

A. Its saturation and cutoff regions
B. Its active region (between the cutoff and saturation regions)
C. Its peak and valley current points
D. Its enhancement and depletion modes

4) Which of the following is an advantage of ceramic capacitors as compared to other types of capacitors?

A. Tight tolerance
B. High stability
C. High capacitance for given volume
D. Comparatively low cost

5) What is meant by the term MMIC?

A. Multi-Megabyte Integrated Circuit
B. Monolithic Microwave Integrated Circuit

C. Military Manufactured Integrated Circuit
D. Mode Modulated Integrated Circuit

6) What is meant when memory is characterized as non-volatile?

A. It is resistant to radiation damage
B. It is resistant to high temperatures
C. The stored information is maintained even if power is removed
D. The stored information cannot be changed once written

7) Which of the following describes a type N connector?

A. A moisture-resistant RF connector useful to 10 GHz
B. A small bayonet connector used for data circuits
C. A threaded connector used for hydraulic systems
D. An audio connector used in surround-sound installations

8) What is the approximate junction threshold voltage of a germanium diode?

A. 0.1 volt
B. 0.3 volts
C. 0.7 volts
D. 1.0 volts

9) Which of these connector types is commonly used for audio signals in Amateur Radio stations?

A. PL-259
B. BNC
C. RCA Phono
D. Type N

10) Which element of a triode vacuum tube is used to regulate the flow of electrons between cathode and plate?

A. Control grid
B. Heater
C. Screen grid
D. Trigger electrode

Answers to this quiz are located in the answer key at the end of the book.

> Any fool can know.
> The point is to
> understand.

\- Albert Einstein

Chapter 7
Practical Circuits

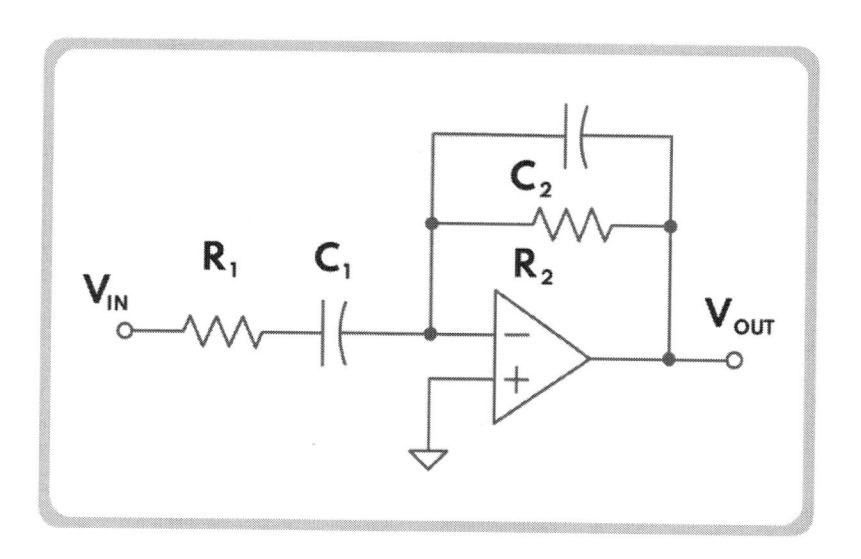

Lesson 7: Practical Circuits

3 Exam questions from this section

Power supplies

Capacitors and inductors are used in a power supply filter network.

Remember how capacitors can store energy? Well, when you power down, you want a bleeder resistor to make sure that the filter capacitors are discharged.

A power supply bleeder resistor **ensures that the filter capacitors are discharged when power is removed.**

The advantage of a switch-mode power supply over a linear power supply is that **high-frequency operation allows the use of small components.**

Rectifiers

180 degrees is the portion of the AC cycle that is converted to DC by a half-wave rectifier.

360 degrees is the portion of the AC cycle that is converted to DC by a full-wave rectifier.

The output wave of an unfiltered full-wave rectifier connected to a resistive load is **a series of DC pulses at twice the frequency of the AC input.**

A full-wave rectifier circuit uses two diodes and a center-tapped transformer.

An advantage of a half-wave rectifier in a power supply is that **only one diode is required.**

Amplifiers

The efficiency of an RF power amplifier is determined as such: **Divide the RF output power by the DC input power.**

Low distortion is a characteristic of a Class A amplifier.

A Class C power stage is appropriate for amplifying a modulated signal in the **FM** mode.

Class C is the class of amplifiers with the highest efficiency.

A linear amplifier is an amplifier in which **the output preserves the input waveform**. Basically, the output will be the same, just smaller or larger.

The reason to neutralize the final amplifier stage of a transmitter is **to eliminate self-oscillations.**

Oscillators

A filter and an amplifier operating in a feedback loop are the basic components of virtually all sine wave oscillators.

The frequency of an LC oscillator is determined by **the inductance and capacitance in the tank circuit.**

A high-stability variable frequency oscillator in a transceiver is a typical application for a direct digital synthesizer (DDS).

Filters

Filters are used to change how circuits respond to different frequencies.

A **filter** is used to process signals from the balanced modulator and then send them to the mixer in some single-sideband phone transmitters.

The impedance of a low-pass filter is **about the same** as the impedance of the transmission line into which it is inserted.

The **cutoff frequency** is the frequency above which a low-pass filter's output power is less than half the input power.

Ultimate rejection specifies a filter's maximum ability to reject signals outside its passband.

The bandwidth of a band-pass filter is measured between **upper and lower half-power** frequencies. **Insertion loss** specifies a filter's attenuation inside its passband.

Digital processing

Sometimes instead of building a complex circuit, it is easy to just use a microcontroller. They are small computers, after all.

A **microcontroller** is an integrated circuit that often can replace complex digital circuitry.

An advantage of using the binary system when processing digital signals is binary "ones" and "zeroes" are easy to represent by an "on" or "off" state.

In a two-input AND gate, **the output is high only when both inputs are high.**

In a two-input NOR gate, **the output is low when either or both inputs are high.**

A 3-bit binary counter has **8** states. You can write all the possible combinations of the 3 bits (1 or 0) like this:

000
001
010
011
100
101
110
111

A shift register is **a clocked array of circuits that passes data in steps along the array.**

Digital signal processor (DSP) filtering is accomplished **by converting the signal from analog to digital and using digital processing.**

Amateur radio continues to advance in digital; it is one of the more exciting parts of the technology.

"Software-defined radio" (SDR) is **a radio in which most major signal processing functions are performed by software.**

Schematic diagram G7-1

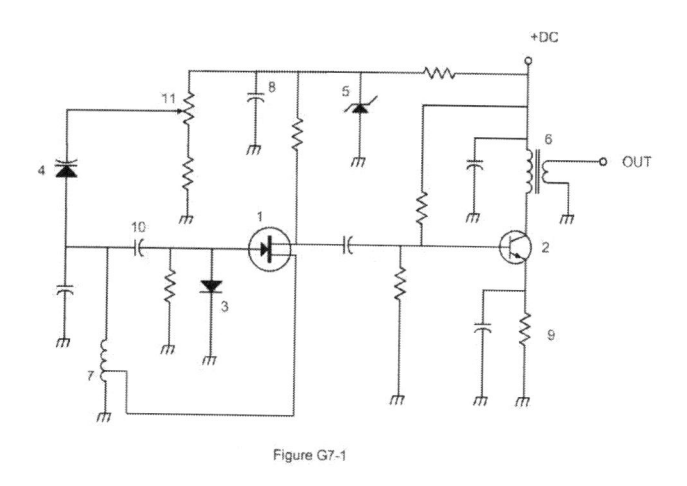

Figure G7-1

Symbol 1: Field effect transistor

Symbol 2: NPN junction transistor

Symbol 5: Zener diode

Symbol 6: Solid Core Transformer

Symbol 7: Tapped inductor

Useful circuits

An advantage of a direct digital synthesizer (DDS) is the **variable frequency with the stability of a crystal oscillator.**

A **discriminator** is used in analog FM receivers to convert IF output signals to audio.

A **mixer** is a circuit used to process signals from the RF amplifier and local oscillator then send the result to the IF filter in a superheterodyne receiver.

A **balanced modulator** is a circuit used to combine signals from the carrier oscillator and speech amplifier and then send the result to the filter in some single-sideband phone transmitters.

90 degrees is the phase difference between the I and Q signals that software-defined radio (SDR) equipment uses for modulation and demodulation.

An advantage of using I and Q signals in software-defined radios (SDRs) is that **all types of modulation can be created with appropriate processing**.

A **product detector** is a circuit used to combine signals from the IF amplifier and BFO (beat frequency oscillator) and send the result to the AF (audio frequency) amplifier in some single-sideband receivers.

The simplest combination of stages that implement a superheterodyne receiver is **HF oscillator, mixer, detector.**

Chapter 7 quiz

1) Which of the following components are used in a power supply filter network?

A. Diodes
B. Transformers and transducers
C. Quartz crystals
D. Capacitors and inductors

2) What portion of the AC cycle is converted to DC by a half-wave rectifier?

A. 90 degrees
B. 180 degrees
C. 270 degrees
D. 360 degrees

3) What is the output waveform of an unfiltered full-wave rectifier connected to a resistive load?

A. A series of DC pulses at twice the frequency of the AC input
B. A series of DC pulses at the same frequency as the AC input
C. A sine wave at half the frequency of the AC input
D. A steady DC voltage

4) Which symbol in figure G7-1 represents a field effect transistor?

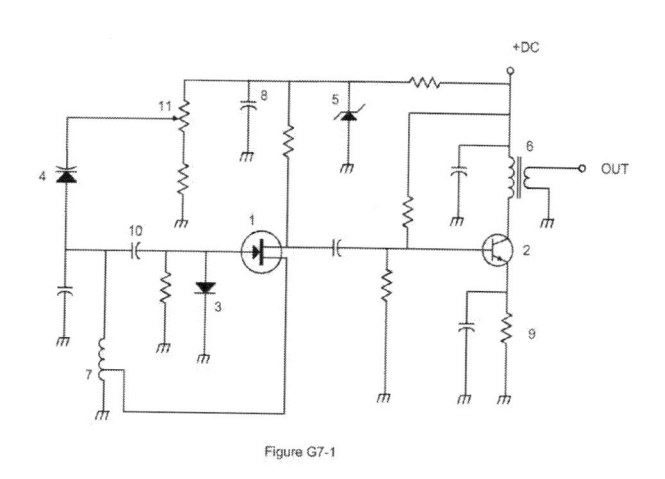

Figure G7-1

A. Symbol 2
B. Symbol 5
C. Symbol 1
D. Symbol 4

5) What is the reason for neutralizing the final amplifier stage of a transmitter?

A. To limit the modulation index
B. To eliminate self-oscillations
C. To cut off the final amplifier during standby periods
D. To keep the carrier on frequency

6) Which of the following are basic components of a sine wave oscillator?
A. An amplifier and a divider
B. A frequency multiplier and a mixer
C. A circulator and a filter operating in a feed-forward loop
D. A filter and an amplifier operating in a feedback loop

7) What determines the frequency of an LC oscillator?

A. The number of stages in the counter
B. The number of stages in the divider
C. The inductance and capacitance in the tank circuit

D. The time delay of the lag circuit

8) Which of the following is used to process signals from the balanced modulator then send them to the mixer in some single sideband phone transmitters?

A. Carrier oscillator
B. Filter
C. IF amplifier
D. RF amplifier

9) The bandwidth of a band-pass filter is measured between what two frequencies?

A. Upper and lower half-power
B. Cutoff and rolloff
C. Pole and zero
D. Image and harmonic

10) What is the phase difference between the I and Q signals that software-defined radio (SDR) equipment uses for modulation and demodulation?

A. Zero
B. 90 degrees
C. 180 degrees
D. 45 degrees

Answers to this quiz are located in the answer key at the end of the book.

> I want to put a ding in the universe.

- Steve Jobs

Chapter 8
Signals and Emissions

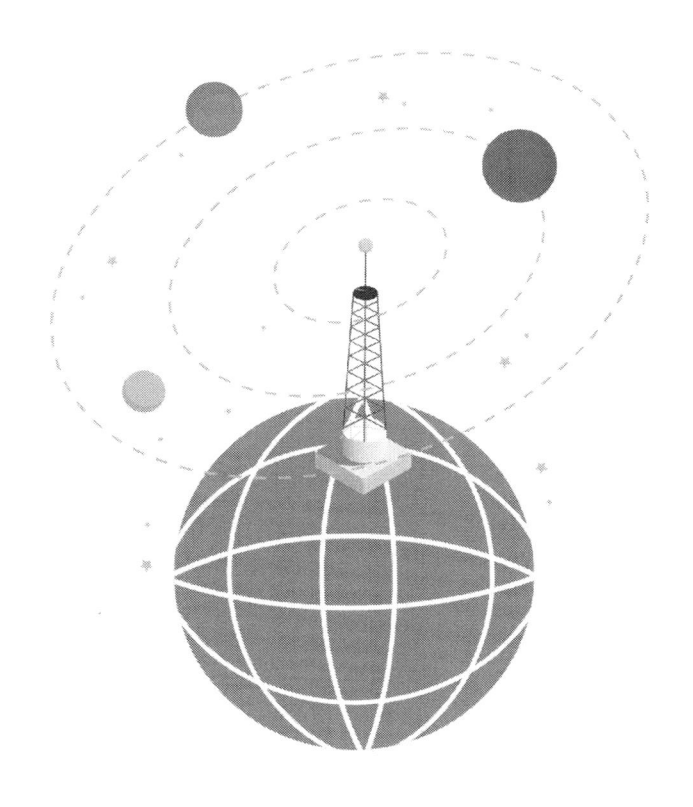

Lesson 8: Signals and Emissions

3 Exam questions from this section

Modulation

Modulation is the key to sending information over ham radio. It is varying some aspect of the radio wave to encode information, so that a receiver on the other end can decode it. There are a few main types of modulation that can be used to send information in ham radio.

Phase modulation is the name of the process that changes the phase angle of an RF wave to convey information. It is produced by a reactance modulator connected to a transmitter RF amplifier stage.

Frequency modulation (FM) is the name of the process that changes the instantaneous frequency of an RF wave to convey information.

Amplitude modulation (AM) varies the instantaneous power level of the RF signal.

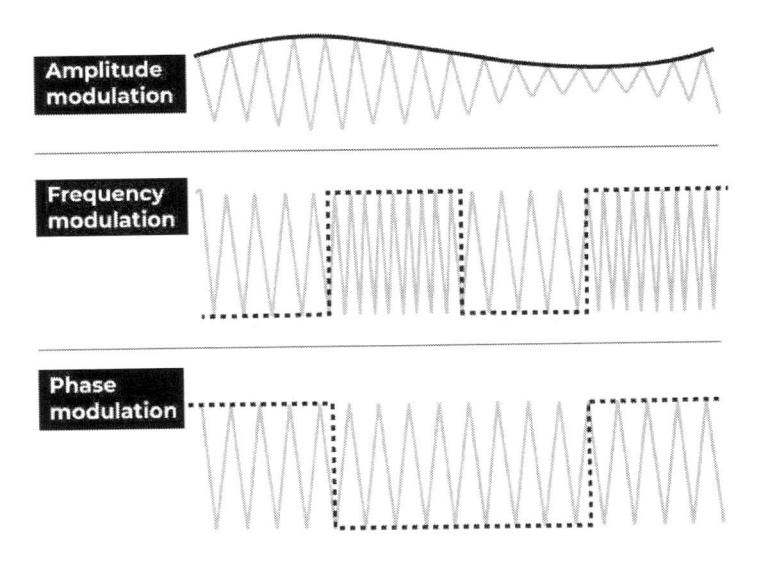

This diagram shows the difference between amplitude, frequency and phase modulation.

One advantage of carrier suppression in a single sideband phone transmission versus full carrier amplitude modulation **is the available** transmitter power can be used more effectively.

The modulation envelope of an AM signal is **the waveform created by connecting the peak values of the modulated signal.**

Bandwidth

The bandwidth of an FM signal is the frequency deviation of the signal, plus the modulating frequencies that carry the signal, times two (two account for both sides of the curve). In simple terms, this is how much "space" the signal takes up.

The formula for FM bandwidth (sideband) is:

$$B_{TOTAL} = (F_{DEV} + F_{MOD}) \times 2$$

For example, **the total bandwidth of an FM phone transmission having 5 kHz** deviation and 3 kHz modulating frequency would be:

(5+3) x 2 = **16 kHz**

You can see from this formula that if you modulate the signal too much (overmodulation), your signal will have more bandwidth and effectively take up more space than needed. Thus, an effect of overmodulation is **excessive bandwidth.**

Single sideband uses the narrowest bandwidth in phone emissions.

Oscillator deviation

Transmitter circuits have a special type of component called an oscillator that creates a varying electrical signal. They then pass this signal to the transmitter. The signal deviates (or changes) to encode information. The ratio of deviation at the oscillator is proportional to the deviation at the transmitter with respect to frequency.

$$\frac{Oscillator\ Deviation}{Oscillator\ Frequency} = \frac{Transmitter\ Deviation}{Transmitter\ Frequency}$$

The test asks us to solve for oscillator deviation given the following values:

- Oscillator frequency = 12.21 MHz
- Transmitter deviation = 5 kHz
- Transmitter frequency = 146.52 MHz

We can adjust the formula to solve for oscillator deviation like this:

$$Oscillator\ Deviation = \frac{(Transmitter\ Deviation)\ x\ (Oscillator\ Frequency)}{Output\ Frequency}$$

Now we need to plug in our numbers. But wait a minute! We need all our frequencies to be in the same units! Let's convert 5 kHz to MHz:

5 kHz = .005 MHz

Now we can plug in our numbers:

$$Oscillator\ Deviation = \frac{(.005\ MHz)\ x\ (12.21\ MHz)}{146.52\ MHz}$$

Time to pull out the calculator. This will leave us with .0004167 MHz. Let's convert that back to Hz (move the decimal by 6 places) to get rid of the extra zeros.

.0004167 MHz = 416.7 Hz

The frequency deviation for a 12.21 MHz reactance-modulated oscillator in a 5 kHz deviation, 146.52 MHz FM phone transmitter is: **416.7 Hz**

Test help: Think 146 is like 416.

Signal distortion

Signal distortion caused by excessive drive is called flat-topping when referring to a single-sideband phone transmission.

Intermodulation is the process that combines two signals in a non-linear circuit or connection to produce unwanted spurious outputs.

The control that is typically used to adjust for proper ALC setting on an amateur single-sideband transceiver is the **transmit audio or microphone gain.**

Electrical concepts

It important to know the duty cycle of the mode you are using when transmitting because **some modes have high duty cycles that could exceed the transmitter's average power rating.**

It is good to match receiver bandwidth to the bandwidth of the operating mode because **it results in the best signal-to-noise ratio.**

Higher symbol rates require wider bandwidth between the transmitted symbol rate and bandwidth.

Devices

The **multiplier** is the stage in a VHF FM transmitter that generates a harmonic of a lower frequency signal to reach the desired operating frequency.

If a receiver mixes a 13.800 MHz VFO with a 14.255 MHz received signal to produce a 455-kHz intermediate frequency (IF) signal, a 13.345 MHz signal will produce an **image response** in the receiver.

Heterodyning is another term for the mixing of two RF signals.

A **mixer** combines a 14.250 MHz input signal with a 13.795 MHz oscillator signal to produce a 455-kHz intermediate frequency (IF) signal.

Digital modes

JT9 and JT65 are modes that are designed to operate at extremely low signal strength on the HF bands.

The part of a data packet containing the routing and handling information is called a **header**.

A 5-bit code with additional start and stop bits is Baudot code.

A receiving station that responds to an ARQ data mode packet containing errors **requests the packet be retransmitted.** Forward error correction (FEC) allows the receiver to correct errors in received data packets **by transmitting redundant information with the data.**

Test help: ARQ is an abbreviation for "automatic repeat request." In AMTOR, an ARQ is sent by the receiving station back to the transmitting station to *request retransmission of missing or corrupted parts of the message.*

8-tone frequency shift keying is a type of modulation used by the FT8 digital mode.

FT8 is a narrow-band digital mode that can receive signals with very low signal-to-noise ratio.

2.4 GHz is the band that shares channels with both the unlicensed Wi-Fi service and the Amateur Radio Service.

A waterfall display shows **frequency is horizontal, signal strength is intensity, time is vertical.** **Overmodulation** is indicated on a waterfall display by one or more vertical lines on either side of a digital signal.

WSPR is a digital mode that is used as a low-power beacon for assessing HF propagation.

FSK

The two separate frequencies of a frequency-shift keyed (FSK) signal are identified by **mark and space.**

An FSK signal is generated **by changing an oscillator's frequency directly with a digital control signal.**

PACTOR

The approximate bandwidth of a PACTOR-III signal at maximum data rate is **2300 Hz.**

In the PACTOR protocol, a NAK response to a transmitted packet means **the receiver is requesting the packet be retransmitted. The connection is dropped** from a failure to exchange information because of excessive transmission attempts when using PACTOR or WINMOR.

PSK31

The type of code used for sending characters in a PSK31 signal is called **Varicode**. In PSK31, **upper case letters use longer Varicode bit sequences and thus slow down transmission.**

The number 31 in "PSK31" refers to **the approximate transmitted symbol rate.**

A characteristic of QPSK31 is:

- It is sideband sensitive.
- Its encoding provides error correction.
- Its bandwidth is approximately the same as BPSK31.

All of the above!

Oscillators

A local oscillator is a mixer input that is varied, or tuned, to convert signals of different frequencies to an intermediate frequency (IF).

The sum and difference combination of a mixer's local oscillator (LO) and RF input frequencies are found in the output.

Chapter 8 quiz

1) Which of the following phone emissions uses the narrowest bandwidth?

A. Single sideband
B. Double sideband
C. Phase modulation
D. Frequency modulation

2) What is the name of the process that changes the instantaneous frequency of an RF wave to convey information?

A. Frequency convolution
B. Frequency transformation
C. Frequency conversion
D. Frequency modulation

3) What emission is produced by a reactance modulator connected to a transmitter RF amplifier stage?

A. Multiplex modulation
B. Phase modulation
C. Amplitude modulation
D. Pulse modulation

4) What is the modulation envelope of an AM signal?

A. The waveform created by connecting the peak values of the modulated signal
B. The carrier frequency that contains the signal
C. Spurious signals that envelop nearby frequencies
D. The bandwidth of the modulated signal

5) If a receiver mixes a 13.800 MHz VFO with a 14.255 MHz received signal to produce a 455 kHz intermediate frequency (IF) signal, what type of interference will a 13.345 MHz signal produce in the receiver?

A. Quadrature noise
B. Image response
C. Mixer interference
D. Intermediate interference

6) What is the total bandwidth of an FM phone transmission having 5 kHz deviation and 3 kHz modulating frequency?

A. 3 kHz
B. 5 kHz
C. 8 kHz
D. 16 kHz

7) What combination of a mixer's Local Oscillator (LO) and RF input frequencies is found in the output?

A. The ratio
B. The average
C. The sum and difference
D. The arithmetic product

8) Which of the following describes Baudot code?

A. A 7-bit code with start, stop and parity bits
B. A code using error detection and correction
C. A 5-bit code with additional start and stop bits
D. A code using SELCAL and LISTEN

9) How are the two separate frequencies of a Frequency Shift Keyed (FSK) signal identified?

A. Dot and dash
B. On and off
C. High and low
D. Mark and space

10) Which type of code is used for sending characters in a PSK31 signal?

A. Varicode
B. Viterbi
C. Volumetric
D. Binary

> **Answers to this quiz are located in the answer key at the end of the book.**

> Tell me and I forget. Teach me and I remember. Involve me and I learn.

— Ben Franklin

Chapter 9
Antennas and Feed Lines

Lesson 9: Antennas and Feed Lines

4 Exam questions from this section

Antenna concepts

An advantage of a horizontally-polarized as compared to a vertically-polarized HF antenna is **lower ground reflection losses.** A disadvantage of multiband antennas is that **they have poor harmonic rejection.**

The "main lobe" of a directive antenna is **the direction of maximum radiated field strength from the antenna.**

In referring to antenna gain, **dBi refers to an isotropic antenna, dBd refers to a dipole antenna.**

dBi gain figures are 2.15 dB higher than dBd gain figures when antenna gain is stated in dBi compared to gain stated in dBd for the same antenna.

The primary purpose of antenna traps is to **permit multiband operation.**

One disadvantage of a directly fed random-wire HF antenna is **that you may experience RF burns when touching metal objects in your station.**

The feed-point impedance of an end-fed half-wave antenna is **very high.**

Broadside to the loop is the direction or directions that an electrically small loop (less than 1/3 wavelength in circumference) has nulls in its radiation pattern.

A "screwdriver" mobile antenna adjusts its feed-point impedance **by varying the base loading inductance.**

A directional antenna is the best HF antenna to utilize when you are trying to minimize interference.

Feed Line: Attenuation

As the frequency of the signal it is carrying increases, coaxial cable's **attenuation increases.**

RF feed line loss usually is expressed in **decibels per 100 feet.**

The distance between the centers of the conductors and the radius of the conductors is one factor that determines the characteristic impedance of a parallel conductor antenna feed line.

The typical characteristic impedance of coaxial cables used for antenna feed lines at amateur stations is **50 and 75 ohms.** The characteristic impedance of flat ribbon TV type twin-lead is **300 ohms.**

A difference between feed line impedance and antenna feed-point impedance might cause reflected power at the point where a feed line connects to an antenna.

To prevent standing waves on an antenna feed line, **the antenna feed point impedance must be matched to the characteristic impedance of the feed line.**

The typical characteristic impedance of "window line" parallel transmission line is **450 ohms.**

Feed line: SWR (standing wave ratio)

The interaction between high standing wave ratio (SWR) and transmission line loss is that **if a transmission line is lossy, high SWR will increase the loss.** The effect of transmission line loss on SWR measured at the input to the line is that **the higher the transmission line loss, the more the SWR will read artificially low.**

For this section, whenever you see that a feed line is connecting to a non-reactive load you can simply take the ohms of the higher impedance and divide it by the lesser. For example:

The standing wave ratio that will result when connecting a 50-ohm feed line to a non-reactive load having 200-ohm impedance is **4:1** (200/50)

The standing wave ratio that will result when connecting a 50-ohm feed line to a non-reactive load having 50-ohm impedance is **1:1** (50/50)

The standing wave ratio that will result when connecting a 50-ohm feed line to a non-reactive load having 25-ohm impedance is **2:1** (50/25)

The standing wave ratio that will result when connecting a 50-ohm feed line to a non-reactive load having 10-ohm impedance is **5:1 (50/10)**

The standing wave ratio that will result when connecting a 50-ohm feed line to an antenna that has a purely resistive 300-ohm feed point impedance is **6:1 (300/50)**.

If the SWR on an antenna feed line is 5 to 1, and a matching network at the transmitter end of the feed line is adjusted to 1 to 1 SWR, the resulting SWR on the feed line is **5 to 1**.

Test help: No math required here because the SWR was adjusted to 1:1.

Dipole

The radiation pattern of a dipole antenna in free space in a plane containing the conductor **is a figure eight at right angles to the antenna.**

A common name of a dipole with a single central support is an **inverted V**.

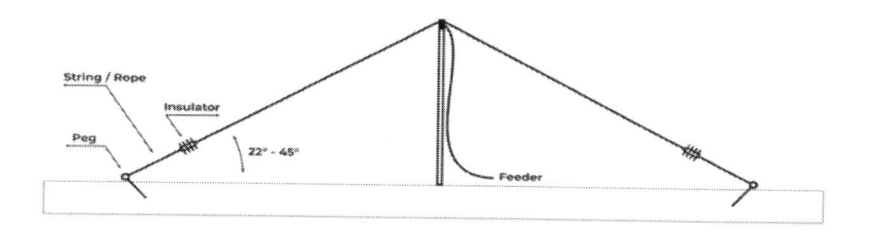

An inverted V antenna is popular on HF bands.

If the antenna is less than 1/2 wavelength high, the azimuthal pattern is almost omnidirectional for the horizontal (azimuthal) radiation pattern of a horizontal dipole HF antenna.

The combined vertical and horizontal polarization pattern of a multi-wavelength, horizontal loop antenna is **virtually omnidirectional with a lower peak vertical radiation angle than a dipole.**

Radial wires of a ground-mounted vertical antenna system are placed **on** the surface of the Earth or buried a few inches below the ground.

The feed point impedance **steadily decreases** when a 1/2-wave dipole antenna is lowered below a 1/4 wave above ground.

The feed point impedance **steadily increases** when a 1/2-wave dipole is moved from the center toward the ends.

To calculate the length of a dipole antenna, simply divide the value of 468 by the frequency in MHz and this will give you the length needed in feet. To find it for 1/4 wavelength, simply use 234.

Examples:

To find the length of a 1/2-wave dipole antenna cut for 14.250 MHz, the formula would be 468/14.250 MHz = **33 feet**

For the length of a 1/2-wave dipole antenna cut for 3.550 MHz, the formula would be 468/3.550 MHz = **132 feet**

For the length of a 1/4-wave dipole antenna cut for 28.5 MHz, the formula would be 234/28.5 MHz = **8 feet**.

Yagi

Larger diameter elements would increase the bandwidth of a Yagi antenna. The approximate length of the driven element of a Yagi antenna is **1/2 wavelength.** A three-element, single-band Yagi antenna's **director is normally the shortest element and the reflector is normally the longest element.**

By increasing boom length and adding directors, a Yagi antenna's **gain increases.**

The reflector element must be approximately 5 percent longer than the driven element when configuring the loops of a two-element quad antenna used to operate as a beam antenna that also assumes that one of the elements is used as a reflector.

The "front-to-back ratio" in Yagi antennas refers to **the power radiated in the major radiation lobe compared to that in the opposite direction.**

An advantage of using a gamma match for impedance matching of a Yagi antenna to 50-ohm coax feed line is that **it does not require that the elements be insulated from the boom.** An advantage of vertical stacking of horizontally polarized Yagi antennas is that **it narrows the main lobe in elevation.**

A shorted transmission line stub placed at the feed point of a Yagi antenna to provide impedance matching is called a beta or hairpin match.

The following are Yagi antenna design variables that could be adjusted to optimize forward gain, front-to-back ratio or SWR bandwidth:

- The physical length of the boom.
- The number of elements on the boom.
- The spacing of each element along the boom.

All of the above!

An advantage of using a gamma match with a Yagi antenna is **that it does not require that the driven element be insulated from the boom.**

The gain of two three-element horizontally polarized Yagi antennas spaced vertically 1/2 wavelength apart typically is **3 dB higher** than the gain of a single three-element Yagi.

The reflector is longer, and the director is shorter on a three-element Yagi reflector and director, compared to that of the driven element.

Quad antennas

The forward gain of a two-element quad antenna is **about the same** compared to the forward gain of a three-element Yagi antenna.

When the feed point of a quad antenna of any shape is moved from the midpoint of the top or bottom to the midpoint of either side, **the polarization of the radiated signal changes from horizontal to vertical.**

Each side of the reflector element of a quad antenna is **slightly more than 1/4 wavelength long**. Each side of the driven element of a quad antenna is approximately **1/4 wavelength long.**

Beverage antennas

A Beverage antenna is **a very long and low directional receiving antenna.** The primary use of a beverage antenna is **directional receiving for low HF bands.**

A Beverage antenna is not used for transmitting because **it has high losses compared to other types of antennas**. An application for a Beverage antenna **is directional receiving for low HF bands**.

Ground plane antennas

Sloping the radials downward is a common way to adjust the feed point impedance of a quarter-wave ground plane vertical antenna to be approximately 50 ohms. When the feed point impedance of a ground plane antenna's radials is changed from horizontal to sloping downward, its feed point impedance **increases**.

Omnidirectional in azimuth best describes the radiation pattern of a quarter-wave, ground-plane vertical antenna.

Log periodic antennas

A log periodic antenna's **element length and spacing vary logarithmically along the boom.**

Wide bandwidth is an advantage of a log periodic antenna.

NVIS antenna

A horizontal dipole placed between 1/10 and 1/4 wavelength above the ground will be most effective as a Near Vertical Incidence Skywave (NVIS) antenna for short-skip communications on 40 meters during the day.

Halo antenna

The maximum radiation from a portable VHF/UHF "halo" antenna is **omnidirectional in the plane of the halo.**

Chapter 9 quiz

1) What are the typical characteristic impedances of coaxial cables used for antenna feed lines at amateur stations?

   A. 25 and 30 ohms
   B. 50 and 75 ohms
   C. 80 and 100 ohms
   D. 500 and 750 ohms

2) What standing wave ratio will result when connecting a 50 ohm feed line to a non-reactive load having 200 ohm impedance?

   A. 4:1
   B. 1:4
   C. 2:1
   D. 1:2

3) How does the attenuation of coaxial cable change as the frequency of the signal it is carrying increases?

A. Attenuation is independent of frequency
B. Attenuation increases
C. Attenuation decreases
D. Attenuation reaches a maximum at approximately 18 MHz

4) Which of the following best describes the radiation pattern of a quarter-wave, ground-plane vertical antenna?

A. Bi-directional in azimuth
B. Isotropic
C. Hemispherical
D. Omnidirectional in azimuth

5) Where should the radial wires of a ground-mounted vertical antenna system be placed?

A. As high as possible above the ground
B. Parallel to the antenna element
C. On the surface of the Earth or buried a few inches below the ground
D. At the center of the antenna

6) What is the approximate length for a 1/4 wave vertical antenna cut for 28.5 MHz?

A. 8 feet
B. 11 feet
C. 16 feet
D. 21 feet

7) What does "front-to-back ratio" mean in reference to a Yagi antenna?

A. The number of directors versus the number of reflectors
B. The relative position of the driven element with respect to the reflectors and directors
C. The power radiated in the major radiation lobe compared to that in the opposite direction
D. The ratio of forward gain to dipole gain

8) How does antenna gain stated in dBi compare to gain stated in dBd for the same antenna?

A. dBi gain figures are 2.15 dB lower than dBd gain figures
B. dBi gain figures are 2.15 dB higher than dBd gain figures
C. dBi gain figures are the same as the square root of dBd gain figures multiplied by 2.15
D. dBi gain figures are the reciprocal of dBd gain figures + 2.15 dB

9) How does the forward gain of a two-element quad antenna compare to the forward gain of a three-element Yagi antenna?

A. About the same
B. About 2/3 as much
C. About 1.5 times as much
D. About twice as much

10) Which of the following is a disadvantage of multiband antennas?

A. They present low impedance on all design frequencies
B. They must be used with an antenna tuner
C. They must be fed with open wire line
D. They have poor harmonic rejection

> **Answers to this quiz are located in the answer key at the end of the book.**

> The man who does not read has no advantage over the man who cannot read.

— Mark Twain

Chapter 10
Electrical and RF Safety

Lesson 10: Electrical Safety

② Exam questions from this section

Electrical shock

To ensure that hazardous voltages cannot appear on the chassis, every item of station equipment should be grounded.

Good practice for lightning protection grounds is **they must be bonded together with all other grounds.**

Current flowing from one or more of the voltage-carrying wires directly to ground will cause a Ground Fault Circuit Interrupter (GFCI) to disconnect the 120- or 240-volt AC line power to a device.

When installing a ground-mounted antenna, **it should be installed such** that it is protected against unauthorized access.

A 15-ampere fuse or circuit breaker would be appropriate to use with a circuit that uses AWG number 14 wiring. The minimum wire size that may be safely used for a circuit that draws up to 20 amperes of continuous current is **AWG number 12.**

Only the two wires carrying voltage in a four-conductor connection should be attached to fuses or circuit breakers in a device operated from a 240-VAC single phase source.

The National Electrical Code covers **electrical safety inside the ham shack**.

According to the National Electrical Code, the minimum wire size that may be used safely for wiring with a 20-ampere circuit breaker is **AWG number 12.**

The purpose of a power supply interlock is **to ensure that dangerous voltages are removed if the cabinet is opened.**

Generators

The **danger of carbon monoxide poisoning** is the primary reason for not placing a gasoline-fueled generator inside an occupied area.

When powering your house from an emergency generator, **you must disconnect the incoming utility power feed. The generator should be located in a well-ventilated area** when you're installing your generator for emergencies.

RF exposure

RF energy can affect a human body **by heating body tissue. The total RF exposure averaged over a certain time** is referred to as "time averaging."

The following properties are important in estimating whether an RF signal exceeds the maximum permissible exposure (MPE):

- Its duty cycle.
- Its frequency.
- Its power density.

All of the above!

You determine that your station complies with FCC RF exposure regulations:

- By calculation based on FCC OET Bulletin 65.
- By calculation based on computer modeling.
- By measurement of field strength using calibrated equipment.

All of the above!

If an evaluation of your station shows that RF energy radiated from your station exceeds permissible limits, you should **take action to prevent human exposure to the excessive RF fields. A lower transmitter duty cycle permits greater short-term exposure levels** when evaluating RF exposure. To ensure compliance with RF safety regulations when transmitter power exceeds levels specified in FCC Part 97.13, you must **perform a routine RF exposure evaluation.**

One thing that can be done if an evaluation shows that a neighbor might receive more than the allowable limit of RF exposure from the main lobe of a directional antenna would be to **take precautions to ensure that the antenna cannot be pointed in their direction.**

One precaution you should you take if you install an indoor transmitting antenna is to **make sure that MPE limits are not exceeded in occupied areas**. A precaution that you should take whenever you make adjustments or repairs to an antenna is to **turn off the transmitter and disconnect the feed line**.

Test help: MPE stands for maximum permissible exposure.

A calibrated field strength meter with a calibrated antenna can be used to accurately measure an RF field.

Soldering

A soldered joint will likely be destroyed by the heat of a lightning strike is the reason why soldered joints should not be used with the wires that connect the base of a tower to a system of ground rods. Lead-tin solder can be a danger because **lead can contaminate food if hands are not washed carefully after handling the solder**.

Tower hazards

When climbing a tower using a safety belt or harness, you should **confirm that the belt is rated for the weight of the climber and that it is within its allowable service life.** Any person preparing to climb a tower that supports electrically powered devices should **make sure all circuits that supply power to the tower are locked out and tagged.**

Chapter 10 quiz

1) What is one way that RF energy can affect human body tissue?

A. It heats body tissue
B. It causes radiation poisoning
C. It causes the blood count to reach a dangerously low level
D. It cools body tissue

2) Which of the following properties is important in estimating whether an RF signal exceeds the maximum permissible exposure (MPE)?

A. Its duty cycle
B. Its frequency
C. Its power density
D. All these choices are correct

3) Which of the following steps must an amateur operator take to ensure compliance with RF safety regulations when transmitter power exceeds levels specified in FCC Part 97.13?

A. Post a copy of FCC Part 97.13 in the station
B. Post a copy of OET Bulletin 65 in the station
C. Perform a routine RF exposure evaluation
D. Contact the FCC for a visit to conduct a station evaluation

4) How can you determine that your station complies with FCC RF exposure regulations?

A. By calculation based on FCC OET Bulletin 65
B. By calculation based on computer modeling
C. By measurement of field strength using calibrated equipment
D. All these choices are correct

5) What type of instrument can be used to accurately measure an RF field?

A. A receiver with an S meter
B. A calibrated field strength meter with a calibrated antenna
C. An SWR meter with a peak-reading function
D. An oscilloscope with a high-stability crystal marker generator

6) What is one thing that can be done if evaluation shows that a neighbor might receive more than the allowable limit of RF exposure from the main lobe of a directional antenna?

A. Change to a non-polarized antenna with higher gain
B. Post a warning sign that is clearly visible to the neighbor
C. Use an antenna with a higher front-to-back ratio
D. Take precautions to ensure that the antenna cannot be pointed in their direction

7) What effect does transmitter duty cycle have when evaluating RF exposure?

A. A lower transmitter duty cycle permits greater short-term exposure levels

B. A higher transmission duty cycle permits greater short-term exposure levels

C. Low duty cycle transmitters are exempt from RF exposure evaluation requirements

D. High duty cycle transmitters are exempt from RF exposure requirements

8) Which of the following conditions will cause a Ground Fault Circuit interrupter (GFCI) to disconnect the 120 or 240 Volt AC line power to a device?

A. Current flowing from one or more of the voltage-carrying wires to the neutral wire
B. Current flowing from one or more of the voltage-carrying wires directly to ground
C. Overvoltage on the voltage-carrying wires
D. All these choices are correct

9) What precaution should you take if you install an indoor transmitting antenna?

A. Locate the antenna close to your operating position to minimize feed-line radiation
B. Position the antenna along the edge of a wall to reduce parasitic radiation

C. Make sure that MPE limits are not exceeded in occupied areas
D. Make sure the antenna is properly shielded

10) What must you do if an evaluation of your station shows RF energy radiated from your station exceeds permissible limits?

A. Take action to prevent human exposure to the excessive RF fields
B. File an Environmental Impact Statement (EIS-97) with the FCC
C. Secure written permission from your neighbors to operate above the controlled MPE limits
D. All these choices are correct

> **Answers to this quiz are located in the answer key at the end of the book.**

Practice Exams

Practice Exam 1

1) Which of the following frequencies is in the General class portion of the 40-meter band in ITU Region 2?

A. 7.250 MHz
B. 7.500 MHZ
C. 40.200 MHz
D. 40.500 MHz

2) What is the maximum height above ground to which an antenna structure may be erected without requiring notification to the FAA and registration with the FCC, provided it is not at or near a public use airport?

A. 50 feet
B. 100 feet
C. 200 feet
D. 300 feet

3) What is the maximum transmitting power an amateur station may use on the 12-meter band?

A. 50 watts PEP output
B. 200 watts PEP output
C. 1500 watts PEP output
D. An effective radiated power equivalent to 100 watts from a half-wave dipole

4) Which of the following must a person have before they can be an administering VE for a Technician class license examination?

A. Notification to the FCC that you want to give an examination
B. Receipt of a Certificate of Successful Completion of Examination (CSCE) for General class
C. Possession of a properly obtained telegraphy license
D. An FCC General class or higher license and VEC accreditation

5) Which of the following would disqualify a third party from participating in stating a message over an amateur station?

A. The third party's amateur license has been revoked and not reinstated
B. The third party is not a U.S. citizen
C. The third party is a licensed amateur
D. The third party is speaking in a language other than English

6) Which of the following modes is most commonly used for voice communications on the 160-meter, 75-meter and 40-meter bands?

A. Upper sideband
B. Lower sideband
C. Vestigial sideband
D. Double sideband

7) Which of the following is true concerning access to frequencies?

A. Nets always have priority
B. QSOs in progress always have priority
C. Except during emergencies, no amateur station has priority access to any frequency
D. Contest operations must always yield to non-contest use of frequencies

8) What should you do if a CW station sends "QRS?"

A. Send slower
B. Change frequency
C. Increase your power
D. Repeat everything twice

9) Which of the following are objectives of the Volunteer Monitoring Program?

A. To conduct efficient and orderly amateur licensing examinations
B. To encourage amateur radio operators to self-regulate and comply with the rules
C. To coordinate repeaters for efficient and orderly spectrum usage
D. To provide emergency and public safety communications

10) Which mode is normally used when sending RTTY signals via AFSK with an SSB transmitter?

A. USB
B. DSB
C. CW
D. LSB

11) Approximately how long does it take the increased ultraviolet and X-ray radiation from solar flares to affect radio propagation on Earth?

A. 28 days
B. 1 to 2 hours
C. 8 minutes
D. 20 to 40 hours

12) What factors affect the MUF?

A. Path distance and location
B. Time of day and season
C. Solar radiation and ionospheric disturbances
D. All these choices are correct

13) What does the term "critical angle" mean, as used in radio wave propagation?

A. The long path azimuth of a distant station
B. The short path azimuth of a distant station
C. The lowest takeoff angle that will return a radio wave to Earth under specific ionospheric conditions
D. The highest takeoff angle that will return a radio wave to Earth under specific ionospheric conditions

14) What is the purpose of the "notch filter" found on many HF transceivers?

A. To restrict the transmitter voice bandwidth
B. To reduce interference from carriers in the receiver passband
C. To eliminate receiver interference from impulse noise sources
D. To enhance the reception of a specific frequency on a crowded band

15) Which of the following is an advantage of an oscilloscope versus a digital voltmeter?

A. An oscilloscope uses less power
B. Complex impedances can be easily measured
C. Input impedance is much lower
D. Complex waveforms can be measured

16) Which of the following could be a cause of interference covering a wide range of frequencies?

A. Not using a balun or line isolator to feed balanced antennas
B. Lack of rectification of the transmitter's signal in power conductors
C. Arcing at a poor electrical connection
D. Using a balun to feed an unbalanced antenna

17) What is the purpose of a speech processor as used in a modern transceiver?

A. Increase the intelligibility of transmitted phone signals during poor conditions
B. Increase transmitter bass response for more natural-sounding SSB signals
C. Prevent distortion of voice signals
D. Decrease high-frequency voice output to prevent out-of-band operation

18) What is the purpose of a capacitance hat on a mobile antenna?

A. To increase the power handling capacity of a whip antenna
B. To allow automatic band changing
C. To electrically lengthen a physically short antenna
D. To allow remote tuning

19) What is impedance?

A. The electric charge stored by a capacitor
B. The inverse of resistance
C. The opposition to the flow of current in an AC circuit
D. The force of repulsion between two similar electric fields

20) How does the total current relate to the individual currents in each branch of a purely resistive parallel circuit?

A. It equals the average of each branch current
B. It decreases as more parallel branches are added to the circuit
C. It equals the sum of the currents through each branch
D. It is the sum of the reciprocal of each individual voltage drop

21) What happens if a signal is applied to the secondary winding of a 4:1 voltage step-down transformer instead of the primary winding?

A. The output voltage is multiplied by 4
B. The output voltage is divided by 4
C. Additional resistance must be added in series with the primary to prevent overload
D. Additional resistance must be added in parallel with the secondary to prevent overload

22) What is an advantage of the low internal resistance of nickel-cadmium batteries?

A. Long life
B. High discharge current
C. High voltage
D. Rapid recharge

23) What determines the performance of a ferrite core at different frequencies?

A. Its conductivity
B. Its thickness
C. The composition, or "mix," of materials used
D. The ratio of outer diameter to inner diameter

24) What useful feature does a power supply bleeder resistor provide?

A. It acts as a fuse for excess voltage
B. It ensures that the filter capacitors are discharged when power is removed
C. It removes shock hazards from the induction coils
D. It eliminates ground loop current

25) Which of these classes of amplifiers has the highest efficiency?

A. Class A
B. Class B
C. Class AB
D. Class C

26) What circuit is used to process signals from the RF amplifier and local oscillator then send the result to the IF filter in a superheterodyne receiver?

A. Balanced modulator
B. IF amplifier
C. Mixer
D. Detector

27) How is an FSK signal generated?

A. By keying an FM transmitter with a sub-audible tone
B. By changing an oscillator's frequency directly with a digital control signal
C. By using a transceiver's computer data interface protocol to change frequencies
D. By reconfiguring the CW keying input to act as a tone generator

28) Which mixer input is varied or tuned to convert signals of different frequencies to an intermediate frequency (IF)?

A. Image frequency
B. Local oscillator
C. RF input
D. Beat frequency oscillator

29) On what band do amateurs share channels with the unlicensed Wi-Fi service?

A. 432 MHz
B. 902 MHz
C. 2.4 GHz
D. 10.7 GHz

30) What is the typical characteristic impedance of "window line" parallel transmission line?

A. 50 ohms
B. 75 ohms
C. 100 ohms
D. 450 ohms

31) Which of the following is a common way to adjust the feed-point impedance of a quarter-wave ground-plane vertical antenna to be approximately 50 ohms?

A. Slope the radials upward
B. Slope the radials downward
C. Lengthen the radials
D. Shorten the radials

32) Which of the following would increase the bandwidth of a Yagi antenna?

A. Larger-diameter elements
B. Closer element spacing
C. Loading coils in series with the element
D. Tapered-diameter elements

33) What is the feed-point impedance of an end-fed half-wave antenna?

A. Very low
B. Approximately 50 ohms
C. Approximately 300 ohms
D. Very high

34) What does "time averaging" mean in reference to RF radiation exposure?

A. The average amount of power developed by the transmitter over a specific 24-hour period
B. The average time it takes RF radiation to have any long-term effect on the body
C. The total time of the exposure
D. The total RF exposure averaged over a certain time

35) Which wire or wires in a four-conductor connection should be attached to fuses or circuit breakers in a device operated from a 240 VAC single phase source?

A. Only the two wires carrying voltage
B. Only the neutral wire
C. Only the ground wire
D. All wires

Practice Exam 2

1) Which of the following frequencies is within the General class portion of the 20-meter phone band?

A. 14005 kHz
B. 14105 kHz
C. 14305 kHz
D. 14405 kHz

2) Which of the following is a purpose of a beacon station as identified in the FCC rules?

A. Observation of propagation and reception
B. Automatic identification of repeaters
C. Transmission of bulletins of general interest to Amateur Radio licensees
D. Identifying net frequencies

3) What must be done before using a new digital protocol on the air?

A. Type-certify equipment to FCC standards
B. Obtain an experimental license from the FCC
C. Publicly document the technical characteristics of the protocol
D. Submit a rule-making proposal to the FCC describing the codes and methods of the technique

4) When must you add the special identifier "AG" after your call sign if you are a Technician class licensee and have a Certificate of Successful Completion of Examination (CSCE) for General class operator privileges, but the FCC has not yet posted your upgrade on its website?

A. Whenever you operate using General class frequency privileges
B. Whenever you operate on any amateur frequency
C. Whenever you operate using Technician frequency privileges
D. A special identifier is not required if your General class license application has been filed with the FCC

5) The frequency allocations of which ITU region apply to radio amateurs operating in North and South America?

A. Region 4
B. Region 3
C. Region 2
D. Region 1

6) Which mode of voice communication is most commonly used on the HF amateur bands?

A. Frequency modulation
B. Double sideband
C. Single sideband
D. Phase modulation

7) What is a practical way to avoid harmful interference on an apparently clear frequency before calling CQ on CW or phone?

A. Send "QRL?" on CW, followed by your call sign; or, if using phone, ask if the frequency is in use, followed by your call sign
B. Listen for 2 minutes before calling CQ
C. Send the letter "V" in Morse code several times and listen for a response, or say "test" several times and listen for a response
D. Send "QSY" on CW or if using phone, announce "the frequency is in use," then give your call sign and listen for a response

8) What does the Q signal "QRL?" mean?

A. "Will you keep the frequency clear?"
B. "Are you operating full break-in?" or "Can you operate full break-in?"
C. "Are you listening only for a specific station?"
D. "Are you busy?" or "Is this frequency in use?"

9) Which of the following is a good way to indicate on a clear frequency in the HF phone bands that you are looking for a contact with any station?

A. Sign your call sign once, followed by the words "listening for a call" -- if no answer, change frequency and repeat
B. Say "QTC" followed by "this is" and your call sign -- if no answer, change frequency and repeat
C. Repeat "CQ" a few times, followed by "this is," then your call sign a few times, then pause to listen, repeat as necessary
D. Transmit an unmodulated carrier for approximately 10 seconds, followed by "this is" and your call sign, and pause to listen -- repeat as necessary

10) What segment of the 80-meter band is most commonly used for digital transmissions?

A. 3570-3600 kHz
B. 3500-3525 kHz
C. 3700-3750 kHz
D. 3775-3825 kHz

11) What is the solar flux index?

A. A measure of the highest frequency that is useful for ionospheric propagation between two points on Earth
B. A count of sunspots that is adjusted for solar emissions
C. Another name for the American sunspot number
D. A measure of solar radiation at 10.7 centimeters wavelength

12) What usually happens to radio waves with frequencies below the MUF and above the LUF when they are sent into the ionosphere?

A. They are bent back to Earth
B. They pass through the ionosphere
C. They are amplified by interaction with the ionosphere
D. They are bent and trapped in the ionosphere to circle Earth

13) What type of propagation allows signals to be heard in the transmitting station's skip zone?

A. Faraday rotation
B. Scatter
C. Chordal hop
D. Short-path

14) What is normally meant by operating a transceiver in "split" mode?

A. The radio is operating at half power
B. The transceiver is operating from an external power source
C. The transceiver is set to different transmit and receive frequencies
D. The transceiver is emitting an SSB signal, as opposed to DSB operation

15) Which of the following is the best instrument to use when checking the keying waveform of a CW transmitter?

A. An oscilloscope
B. A field strength meter
C. A sidetone monitor
D. A wavemeter

16) What is the effect on an audio device when there is interference from a nearby CW transmitter?

A. On-and-off humming or clicking
B. A CW signal at a nearly pure audio frequency
C. A chirpy CW signal
D. Severely distorted audio

17) How does a signal that reads 20 dB over S9 compare to one that reads S9 on a receiver, assuming a properly calibrated S meter?

A. It is 10 times less powerful
B. It is 20 times less powerful
C. It is 20 times more powerful
D. It is 100 times more powerful

18) What is the name of the process by which sunlight is changed directly into electricity?

A. Photovoltaic conversion
B. Photon emission
C. Photosynthesis
D. Photon decomposition

19) What is reactance?

A. Opposition to the flow of direct current caused by resistance
B. Opposition to the flow of alternating current caused by capacitance or inductance
C. A property of ideal resistors in AC circuits
D. A large spark produced at switch contacts when an inductor is de-energized

20) How many watts of electrical power are used by a 12 VDC light bulb that draws 0.2 amperes?

A. 2.4 watts
B. 24 watts
C. 6 watts
D. 60 watts

21) What is the equivalent capacitance of two 5.0 nanofarad capacitors and one 750 picofarad capacitor connected in parallel?

A. 576.9 nanofarads
B. 1733 picofarads
C. 3583 picofarads
D. 10.750 nanofarads

22) What happens when an inductor is operated above its self-resonant frequency?

A. Its reactance increases
B. Harmonics are generated
C. It becomes capacitive
D. Catastrophic failure is likely

23) How is an LED biased when emitting light?

A. Beyond cutoff
B. At the Zener voltage
C. Reverse biased
D. Forward biased

24) What is an advantage of a half-wave rectifier in a power supply?

A. Only one diode is required
B. The ripple frequency is twice that of a full-wave rectifier
C. More current can be drawn from the half-wave rectifier
D. The output voltage is two times the peak output voltage of the transformer

25) How many states does a 3-bit binary counter have?

A. 3
B. 6
C. 8
D. 16

26) What should be the impedance of a low-pass filter as compared to the impedance of the transmission line into which it is inserted?

A. Substantially higher
B. About the same
C. Substantially lower
D. Twice the transmission line impedance

27) Which of the following is an effect of overmodulation?

A. Insufficient audio
B. Insufficient bandwidth
C. Frequency drift
D. Excessive bandwidth

28) Why is it important to know the duty cycle of the mode you are using when transmitting?

A. To aid in tuning your transmitter
B. Some modes have high duty cycles that could exceed the transmitter's average power rating
C. To allow time for the other station to break in during a transmission
D. The attenuator will have to be adjusted accordingly

29) In the PACTOR protocol, what is meant by a NAK response to a transmitted packet?

A. The receiver is requesting the packet be retransmitted
B. The receiver is reporting the packet was received without error
C. The receiver is busy decoding the packet
D. The entire file has been received correctly

30) If the SWR on an antenna feed line is 5 to 1, and a matching network at the transmitter end of the feed line is adjusted to 1 to 1 SWR, what is the resulting SWR on the feed line?

A. 1 to 1
B. 5 to 1
C. Between 1 to 1 and 5 to 1 depending on the characteristic impedance of the line
D. Between 1 to 1 and 5 to 1 depending on the reflected power at the transmitter

31) How does antenna height affect the horizontal (azimuthal) radiation pattern of a horizontal dipole HF antenna?

A. If the antenna is too high, the pattern becomes unpredictable
B. Antenna height has no effect on the pattern
C. If the antenna is less than 1/2 wavelength high, the azimuthal pattern is almost omnidirectional
D. If the antenna is less than 1/2 wavelength high, radiation off the ends of the wire is eliminated

32) What configuration of the loops of a two-element quad antenna must be used for the antenna to operate as a beam antenna, assuming one of the elements is used as a reflector?

A. The driven element must be fed with a balun transformer
B. There must be an open circuit in the driven element at the point opposite the feed point
C. The reflector element must be approximately 5 percent shorter than the driven element
D. The reflector element must be approximately 5 percent longer than the driven element

33) How does a "screwdriver" mobile antenna adjust its feed-point impedance?

A. By varying its body capacitance
B. By varying the base loading inductance
C. By extending and retracting the whip
D. By deploying a capacitance hat

34) What type of instrument can be used to accurately measure an RF field?

A. A receiver with an S meter
B. A calibrated field strength meter with a calibrated antenna
C. An SWR meter with a peak-reading function
D. An oscilloscope with a high stability crystal marker generator

35) Which of these choices should be observed when climbing a tower using a safety belt or harness?

A. Never lean back and rely on the belt alone to support your weight
B. Confirm that the belt is rated for the weight of the climber and that it is within its allowable service life
C. Ensure that all heavy tools are securely fastened to the belt D-ring
D. All these choices are correct

Practice Exam 3

1) What is the appropriate action if, when operating on either the 30-meter or 60-meter bands, a station in the primary service interferes with your contact?

A. Notify the FCC's regional Engineer in Charge of the interference
B. Increase your transmitter's power to overcome the interference
C. Attempt to contact the station and request that it stop the interference
D. Move to a clear frequency or stop transmitting

2) What are the restrictions on the use of abbreviations or procedural signals in the Amateur Service?

A. Only "Q" signals are permitted
B. They may be used if they do not obscure the meaning of a message
C. They are not permitted
D. Only "10 codes" are permitted

3) What is the maximum power limit on the 60-meter band?

A. 1500 watts PEP
B. 10 watts RMS
C. ERP of 100 watts PEP with respect to a dipole
D. ERP of 100 watts PEP with respect to an isotropic antenna

4) Which of the following criteria must be met for a non-U.S. citizen to be an accredited Volunteer Examiner?

A. The person must be a resident of the U.S. for a minimum of 5 years
B. The person must hold an FCC granted Amateur Radio license of General class or above
C. The person's home citizenship must be in ITU region 2
D. None of these choices is correct; a non-U.S. citizen cannot be a Volunteer Examiner

5) In what part of the 13-centimeter band may an amateur station communicate with non-licensed Wi-Fi stations?

A. Anywhere in the band
B. Channels 1 through 4
C. Channels 42 through 45
D. No part

6) Which of the following is an advantage when using single sideband, as compared to other analog voice modes on the HF amateur bands?

A. Very high fidelity voice modulation
B. Less subject to interference from atmospheric static crashes
C. Ease of tuning on receive and immunity from impulse noise
D. Less bandwidth used and greater power efficiency

7) When selecting a CW transmitting frequency, what minimum separation should be used to minimize interference to stations on adjacent frequencies?

A. 5 to 50 Hz
B. 150 to 500 Hz
C. 1 to 3 KHz
D. 3 to 6 kHz

8) What is the best speed to use when answering a CQ in Morse code?

A. The fastest speed at which you are comfortable copying, but no slower than the CQ
B. The fastest speed at which you are comfortable copying, but no faster than the CQ
C. At the standard calling speed of 10 wpm
D. At the standard calling speed of 5 wpm

9) Which of the following are examples of the NATO Phonetic Alphabet?

A. Able, Baker, Charlie, Dog
B. Adam, Boy, Charles, David
C. America, Boston, Canada, Denmark
D. Alpha, Bravo, Charlie, Delta

10) Which of the following is a requirement when using the FT8 digital mode?

A. A special hardware modem
B. Computer time accurate within approximately 1 second
C. Receiver attenuator set to -12 dB
D. A vertically polarized antenna

11) How long does it take charged particles from coronal mass ejections to affect radio propagation on Earth?

A. 28 days
B. 14 days
C. 4 to 8 minutes
D. 20 to 40 hours

12) What does MUF stand for?

A. The Minimum Usable Frequency for communications between two points
B. The Maximum Usable Frequency for communications between two points
C. The Minimum Usable Frequency during a 24-hour period
D. The Maximum Usable Frequency during a 24-hour period

13) What is Near Vertical Incidence Skywave (NVIS) propagation?

A. Propagation near the MUF
B. Short distance MF or HF propagation using high elevation angles
C. Long path HF propagation at sunrise and sunset
D. Double hop propagation near the LUF

14) What condition can lead to permanent damage to a solid-state RF power amplifier?

A. Insufficient drive power
B. Low input SWR
C. Shorting the input signal to ground
D. Excessive drive power

15) Which of the following can be determined with a field strength meter?

A. The radiation resistance of an antenna
B. The radiation pattern of an antenna
C. The presence and amount of phase distortion of a transmitter
D. The presence and amount of amplitude distortion of a transmitter

16) How can a ground loop be avoided?

A. Connect all ground conductors in series
B. Connect the AC neutral conductor to the ground wire
C. Avoid using lock washers and star washers when making ground connections
D. Connect all ground conductors to a single point

17) How close to the upper edge of the phone segment should your displayed carrier frequency be when using 3 kHz wide USB?

A. At least 3 kHz above the edge of the band
B. At least 3 kHz below the edge of the band
C. At least 1 kHz above the edge of the segment
D. At least 1 kHz below the edge of the segment

18) What is the reason that a series diode is connected between a solar panel and a storage battery that is being charged by the panel?

A. The diode serves to regulate the charging voltage to prevent overcharge
B. The diode prevents self-discharge of the battery through the panel during times of low or no illumination
C. The diode limits the current flowing from the panel to a safe value
D. The diode greatly increases the efficiency during times of high illumination

19) Which of the following devices can be used for impedance matching at radio frequencies?

A. A transformer
B. A Pi-network
C. A length of transmission line
D. All these choices are correct

20) What is the output PEP of an unmodulated carrier if an average reading wattmeter connected to the transmitter output indicates 1060 watts?

A. 530 watts
B. 1060 watts
C. 1500 watts
D. 2120 watts

21) Why is the conductor of the primary winding of many voltage step-up transformers larger in diameter than the conductor of the secondary winding?

A. To improve the coupling between the primary and secondary
B. To accommodate the higher current of the primary
C. To prevent parasitic oscillations due to resistive losses in the primary
D. To ensure that the volume of the primary winding is equal to the volume of the secondary winding

22) What is the primary purpose of a screen grid in a vacuum tube?

A. To reduce grid-to-plate capacitance
B. To increase efficiency
C. To increase the control grid resistance
D. To decrease plate resistance

23) Which of these connector types is commonly used for RF connections at frequencies up to 150 MHz?

A. Octal
B. RJ-11
C. PL-259
D. DB-25

24) Which symbol in figure G7-1 represents a solid core transformer?

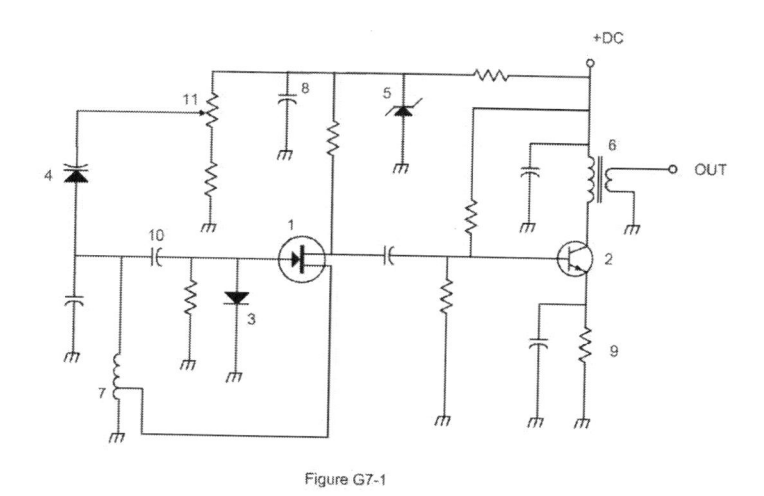

Figure G7-1

A. Symbol 4
B. Symbol 7
C. Symbol 6
D. Symbol 1

25) How is the efficiency of an RF power amplifier determined?

A. Divide the DC input power by the DC output power
B. Divide the RF output power by the DC input power
C. Multiply the RF input power by the reciprocal of the RF output power
D. Add the RF input power to the DC output power

26) What term specifies a filter's maximum ability to reject signals outside its passband?

A. Notch depth
B. Rolloff
C. Insertion loss
D. Ultimate rejection

27) What is meant by the term "flat-topping," when referring to a single sideband phone transmission?

A. Signal distortion caused by insufficient collector current
B. The transmitter's automatic level control (ALC) is properly adjusted
C. Signal distortion caused by excessive drive
D. The transmitter's carrier is properly suppressed

28) Why is it good to match receiver bandwidth to the bandwidth of the operating mode?

A. It is required by FCC rules
B. It minimizes power consumption in the receiver
C. It improves impedance matching of the antenna
D. It results in the best signal-to-noise ratio

29) What is indicated on a waterfall display by one or more vertical lines on either side of a digital signal?

A. Long path propagation
B. Backscatter propagation
C. Insufficient modulation
D. Overmodulation

30) What is the effect of transmission line loss on SWR measured at the input to the line?

A. The higher the transmission line loss, the more the SWR will read artificially low
B. The higher the transmission line loss, the more the SWR with read artificially high
C. The higher the transmission line loss, the more accurate the SWR measurement will be
D. Transmission line loss does not affect the SWR measurement

31) Which of the following is an advantage of a horizontally polarized as compared to a vertically polarized HF antenna?

A. Lower ground reflection losses
B. Lower feed-point impedance
C. Shorter radials
D. Lower radiation resistance

32) Which of the following is an advantage of using a gamma match with a Yagi antenna?

A. It does not require that the driven element be insulated from the boom
B. It does not require any inductors or capacitors
C. It is useful for matching multiband antennas
D. All these choices are correct

33) What is the combined vertical and horizontal polarization pattern of a multi-wavelength, horizontal loop antenna?

A. A figure eight, similar to a dipole
B. Four major loops with deep nulls
C. Virtually omnidirectional with a lower peak vertical radiation angle than a dipole
D. Radiation maximum is straight up

34) What precaution should you take if you install an indoor transmitting antenna?

A. Locate the antenna close to your operating position to minimize feed-line radiation
B. Position the antenna along the edge of a wall to reduce parasitic radiation
C. Make sure that MPE limits are not exceeded in occupied areas
D. Make sure the antenna is properly shielded

35) Which of the following is good practice for lightning protection grounds?

A. They must be bonded to all buried water and gas lines
B. Bends in ground wires must be made as close as possible to a right angle
C. Lightning grounds must be connected to all ungrounded wiring
D. They must be bonded together with all other ground

Scheduling and taking your exam

You are ready to test when you regularly score *at least* 80 percent or more on your practice exams!

Here's what to expect as you get ready to take your official FCC exam.

For additional free practice tests, visit:
www.HamRadioPrep.com/free-ham-radio-practice-tests/

How to find an exam

You're ready to schedule your exam, and you have options -- in person or online!

For most people, the easiest way to take the exam is in person with a friendly, local ham radio club. Volunteer Examiners, who are FCC accredited, will proctor your exam. Volunteer Examiner Coordinators are the only organizations legally allowed to conduct ham radio license exam sessions in the United States. To make sure you're finding an FCC-accredited group, go here:

HamRadioPrep.com/find-an-exam-near-you/

You can search by ZIP code and find the most up-to-date list.

If an in-person exam doesn't sound right for you, VECs now offer remote and online exams. Keep in mind that you will need access to a reliable internet connection, two different devices to connect to a remote video session, and a secure area to be in while you're taking the exam.

You need to be comfortable operating devices such as a computer, tablet or cell phone to connect. You can contact VECs to line up online examinations at this link:

HamRadioPrep.com/ham-radio-license-test-online/

What to bring on exam day

Whether you test online, or in person, you will need to have the following items with you.

1) A legal photo ID or two forms of identification.
2) Your FCC FRN (Federal Registration Number) or Social Security number. We highly recommend people use an FRN for privacy purposes. Please visit

apps.fcc.gov/coresWeb/publicHome.do to obtain your FRN.
3) A calculator is allowed if you have cleared all memory and formulas.

If testing in person, you will also need:
1) Two No. 2 pencils.
2) A check or money order for payment.

You cannot bring notes or other documents or study materials with you for either form of the exam. If you are taking the exam online, you will not need a check or money order and will be prompted ahead of the exam to pay any associated fees online.

What is the exam like?

Whether you take the exam in person or online there will be testers watching you the whole time. They are charged with upholding the test standards, but they are passionate about ham radio and should not make you nervous. They want you to pass your test, too!

If you're testing online, exam proctors will ask that you keep both hands on your keyboard the entire time you test. They will show you which keys you will use to answer questions and move through the test.

Your cell phone will be used to help the proctors monitor your remote testing environment. They will let you know if you need to change something you're doing to stay within the guidelines during your test.

The General exam has 35 questions (just like the practice exams in this book). You do not earn any additional points for completing the exam quickly; you just need to stay within the time allowed by the test examiners.

Be confident! You've done all the prep work; you're passing your practice exams ...

– YOU'VE GOT THIS!

Unauthorized conduct and test rules

The Federal Communications Commission, which grants amateur radio licenses, takes the testing process seriously and instills a variety of guidelines to ensure the integrity of all exams that are administered.

In order to obtain an FCC-issued amateur radio license, you must do one of two things:

- Appear at a local test session conducted by a local amateur radio club. Local amateur radio clubs are authorized by national Volunteer Examiner Coordinators to conduct testing on a local basis with accredited test session administrators, who are known as Volunteer Examiners.

- Take your test online from a national Volunteer Examiner Coordinator that will monitor your test session remotely with cameras in a room clear of any possible test aids. VECs can be located anywhere within the FCC's jurisdiction in the 50 states or territories, and those taking the exam can be located anywhere in the 50 states or territories, too. It doesn't matter where you or the examiners are located when taking your exam online!

FCC rules state the following for those administering tests:

Each VE must observe those taking their tests throughout the entire exam. VEs are charged with ensuring proper conduct and supervision of each exam. If a VE determines that a person taking an exam is not complying with their instructions, they are required to terminate the exam immediately.

VEs may not administer exams to family members, including his or her spouse, children, grandchildren, stepchildren, parents, grandparents, stepparents, brothers, sisters, stepbrothers, stepsisters, aunts, uncles, nieces, nephews and in-laws.

It goes without saying, but VEs may not administer or certify any exam by fraudulent means, or for any types or bribes or reimbursement in excess of that permitted by FCC rules. Any licensed amateur radio operator found guilty of this may find the FCC revoking their amateur station license and suspending the grant of the VE's amateur operator license.

Also, the FCC states that any compromised exam may not be administered to any examinee and that the same question set may not be readministered to the same examinee.

If the FCC determines that there was an issue with the administration of a test in which a license was issued, the FCC may choose to readminister any exam previously given by VEs. If the person fails to appear for re-administration of an exam, the FCC may cancel the operator/primary station license of the licensee in question. If a person in question already holds a Technician or higher license, the FCC can revoke the higher class license in question for failure to appear for reexamination. Thus, if you are a General license holder and the FCC is challenging your Extra class exam, you can be reverted back to your previous General class license by the commission.

Check our website for more information about the testing process at www.HamRadioPrep.com.

Are you enjoying this course? Please leave us a review on Amazon and receive a free gift on us!

As a token of our appreciation for your business and positive review, we will send you a small gift.

★★★★★ **Easy to follow, quizzes after every chapter!**
Reviewed in the United States on March 24, 2019
Verified Purchase

Super easy to follow! There are quizzes after every chapter with answers at the end of the book. Also there's a free coupon for the online course at Hamradioprep.com. Highly recommend for anyone trying to study for their license and very happy with my purchase!

Gregory K.

★★★★★ **Without a doubt the best study manual on the market!**
Reviewed in the United States on July 14, 2020
Verified Purchase

Without a doubt the best study manual on the market! Not only is this manual extremely helpful in itself but the free access to their online course absolutely guarantees that you can pass your test on the first try.

Jim

★★★★★ **Great tool to earn your license!**
Reviewed in the United States on March 25, 2019

Simple to read and easy way to prep yourself for the ham radio technician's exam. I passed the exam easily after reading the book for preparation. Highly Recommend!

Take a screenshot of your review and go to **www.HamRadioPrep.com/book** to upload the screenshot of your five-star review!

Celebrate your achievement!

You've studied, you're prepared, and we know you're going to pass your exam.

After passing your test, commemorate your hard work!

Order a Ham Radio Prep *Exclusive Wall Plaque* for your office or home, and your new FCC license, call sign and grant dates will be beautifully engraved and on display!

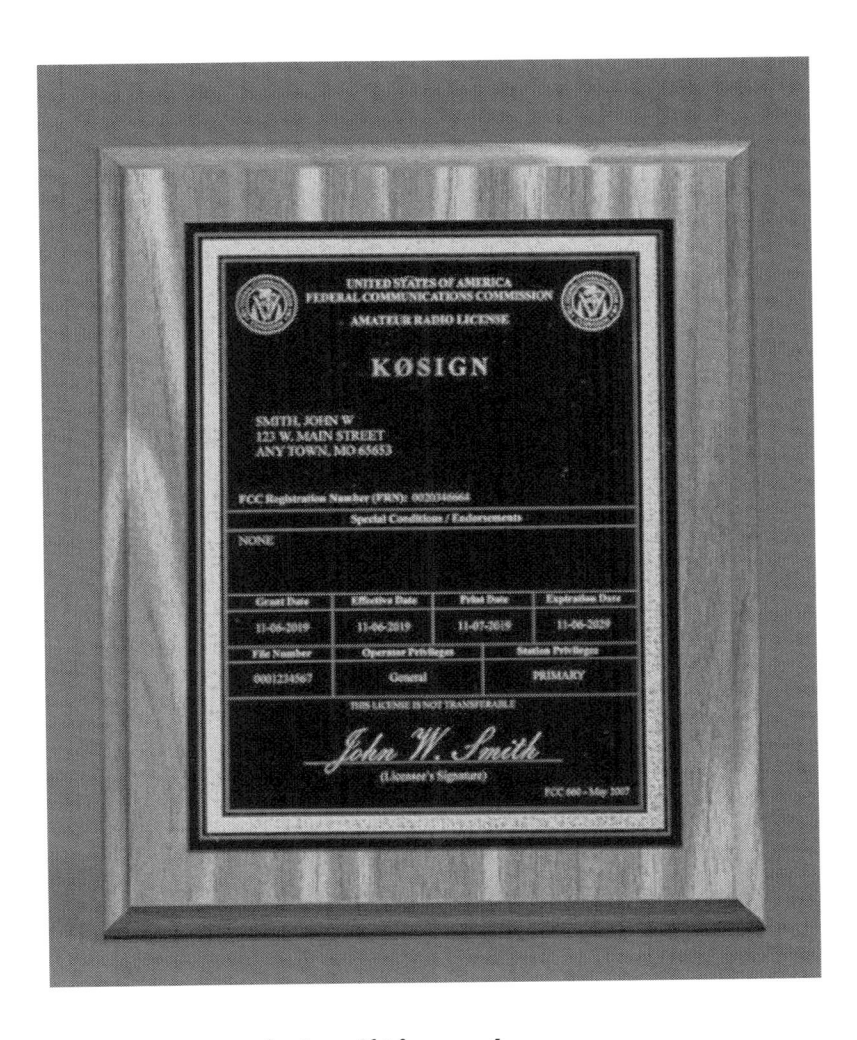

These *special edition plaques* are made here in the USA. They're available in multiple colors and feature a steel metal overlay, exclusively available from Ham Radio Prep. Order here: shop.HamRadioPrep.com

Resources

Our website: www.HamRadioPrep.com

Facebook: www.facebook.com/HamRadioPrep

Reddit: www.reddit.com/user/HamRadioPrep

Twitter: twitter.com/HamPrep

YouTube: www.youtube.com/c/HamRadioPrep

Instagram: www.instagram.com/HamRadioPrep

Scheduling online tests: HamRadioPrep.com/ham-radio-license-test-online

Scheduling in-person tests: HamRadioPrep.com/find-an-exam-near-you

WinLink: www.winlink.org

Finding a repeater: www.repeaterbook.com

Phonetic alphabet:

Phonetic Alphabet

- **A**lpha
- **B**ravo
- **C**harlie
- **D**elta
- **E**cho
- **F**oxtrot
- **G**olf
- **H**otel
- **I**ndia
- **J**uliet
- **K**ilo
- **L**ima
- **M**ike
- **N**ovember
- **O**scar
- **P**apa
- **Q**uebec
- **R**omeo
- **S**ierra
- **T**ango
- **U**niform
- **V**ictor
- **W**hiskey
- **X**ray
- **Y**ankee
- **Z**ulu

HAM RADIO PREP

International Morse Code:

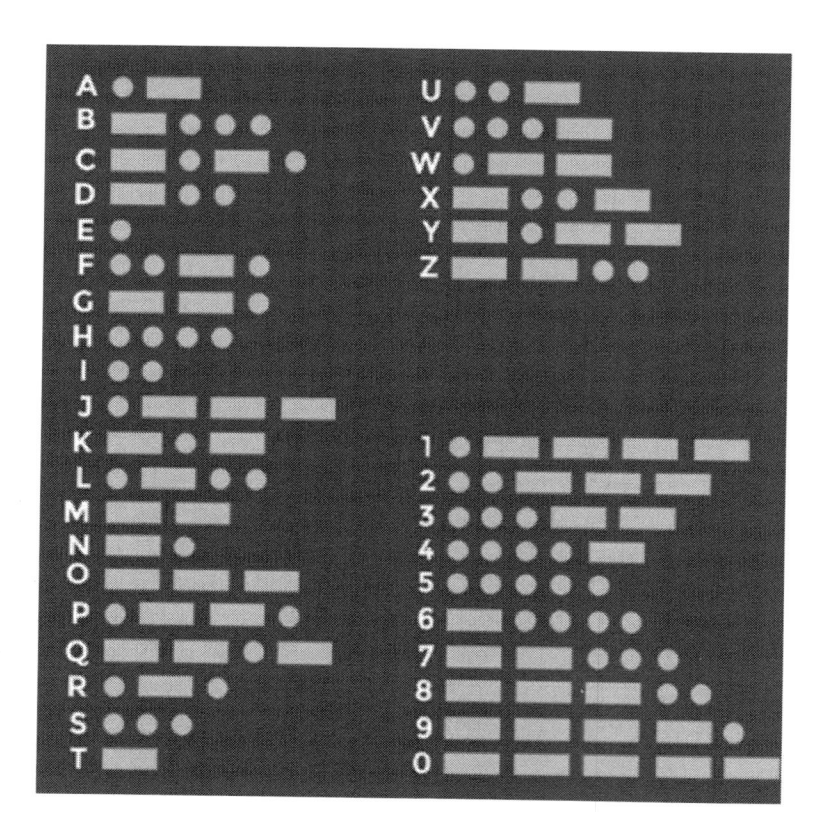

A ·—		U ··—	
B —···		V ···—	
C —·—·		W ·——	
D —··		X —··—	
E ·		Y —·——	
F ··—·		Z ——··	
G ——·			
H ····			
I ··			
J ·———		1 ·————	
K —·—		2 ··———	
L ·—··		3 ···——	
M ——		4 ····—	
N —·		5 ·····	
O ———		6 —····	
P ·——·		7 ——···	
Q ——·—		8 ———··	
R ·—·		9 ————·	
S ···		0 —————	
T —			

GENERAL CLASS
license frequency privileges

LF
1 watt EIRP maximum

2200 meters 135.7 - 137.8 kHz   CW, phone, image, RTTY/data

MF
5 watts EIRP maximum (except 1 watt EIRP in Alaska within 496 miles of Russia)

630 meters 472 - 479 kHz   CW, phone, image, RTTY/data

HF
1,500 watts PEP maximum (unless noted)

160 meters 1.800 - 2.000 MHz   CW, phone, image, RTTY/data
80 meters 3.525 - 3.600 MHz   CW, RTTY/data
               3.800 - 4.000 MHz   CW, phone, image

60 meters
100 watts ERP into antenna with 0 dBd gain

	Five 2.8-kHz channels (center channel shown)
5.332 MHz	CW, phone, narrow digital modes per rules
5.348 MHz	CW, phone, narrow digital modes per rules
5.3585 MHz	CW, phone, narrow digital modes per rules
5.373 MHz	CW, phone, narrow digital modes per rules
5.405 MHz	CW, phone, narrow digital modes per rules

© HAM RADIO PREP

GENERAL CLASS
license frequency privileges

40 meters	7.025	-	7.125 MHz	CW, RTTY/data
	7.125	-	7.300 MHz	CW, phone, image
30 meters				
200 watts ERP maximum				
	10.100	-	10.150 MHz	CW, RTTY/data
20 meters	14.025	-	14.150 MHz	CW, RTTY/data
	14.225	-	14.350 MHz	CW, phone, image
17 meters	18.068	-	18.110 MHz	CW, RTTY/data
	18.110	-	18.168 MHz	CW, phone, image
15 meters	21.025	-	21.200 MHz	CW, RTTY/data
	21.275	-	27.450 MHz	CW, phone, image
12 meters	24.890	-	24.930 MHz	CW, RTTY/data
	24.930	-	24.990 MHz	CW, phone, image
10 meters	28.000	-	28.300 MHz	CW, RTTY/data
	28.300	-	29.700 MHz	CW, phone, image

© HAM RADIO PREP

GENERAL CLASS
license frequency privileges

VHF
1,500 watts PEP maximum

6 meters	50.000	- 50.100 MHz	CW
	50.100	- 54.000 MHz	CW, digital, SSB, AM, FM, TV
2 meters	144.000	- 144.100 MHz	CW
	144.100	- 148.000 MHz	CW, MCW, digital, SSB, AM, FM, TV
1.25 meters	219.000	- 220.000 MHz	Point-to-point digital links - 50 watts PEP maximum; 100 kHz bandwidth
	222.000	- 225.000 MHz	CW, MCW, digital, SSB, AM, FM, TV

UHF
1,500 watts PEP maximum

70 cm	420.000	- 450.000 MHz	CW, MCW, digital, SSB, AM, FM, TV
	420.000	- 430.000 MHz	Not available for use north of Line A near Canada border
33 cm	902.000	- 928.000 MHz	CW, MCW, digital, SSB, AM, FM, TV
23 cm	1.240	- 1.300 GHz	CW, MCW, digital, SSB, AM, FM, TV
13 cm	2.300	- 2.310 GHz	CW, digital, SSB, AM, FM, TV
	2.390	- 2.450 GHz	CW, digital, SSB, AM, FM, TV

© HAM RADIO PREP

GENERAL CLASS
license frequency privileges

SHF
1,500 watts PEP maximum

9 cm	3.300 - 3.500 GHz	CW, digital, SSB, AM, FM, TV
5 cm	5.650 - 5.925 GHz	CW, digital, SSB, AM, FM, TV
3 cm	10.000 - 10.500 GHz	CW, digital, SSB, AM, FM, TV
1.2 cm	24.000 - 24.250 GHz	CW, digital, SSB, AM, FM, TV

EHF
1,500 watts PEP maximum

6 mm	47.000 - 47.200 GHz	CW, digital, SSB, AM, FM, TV
4 mm	76.000 - 81.000 GHz	CW, digital, SSB, AM, FM, TV
2.5 mm	122.250 - 123.000 GHz	CW, digital, SSB, AM, FM, TV
2 mm	134.000 - 141.000 GHz	CW, digital, SSB, AM, FM, TV
1 mm	241.000 - 250.000 GHz	CW, digital, SSB, AM, FM, TV
All above	275 GHz and above	CW, digital, SSB, AM, FM, TV

© HAM RADIO PREP

Answer Keys

Chapter 1: Commission's Rules

1) C   2) D   3) D   4) B
5) A   6) B   7) C   8) D
9) B   10) A

Chapter 2: Operating Procedures

1) A   2) C   3) B   4) D
5) B   6) A   7) B   8) D
9) C   10) D

Chapter 3: Radio Wave Propagation

1) B   2) A   3) C   4) A
5) D   6) A   7) C   8) B
9) B   10) A

Chapter 4: Amateur Radio Practices

1) C 2) A 3) D 4) C
5) B 6) A 7) A 8) D
9) A 10) C

Chapter 5: Electrical Principles

1) D 2) B 3) D 4) A
5) B 6) C 7) C 8) B
9) A 10) B

Chapter 6: Circuit Components

1) C 2) B 3) A 4) D
5) B 6) C 7) A 8) B
9) C 10) A

Chapter 7: Practical Circuits

1) D 2) B 3) A 4) C
5) B 6) D 7) C 8) B
9) A 10) B

Chapter 8: Signals and Emissions

1) A   2) D   3) B   4) A
5) B   6) D   7) C   8) C
9) D   10) A

Chapter 9: Antennas and Feed Lines

1) B   2) A   3) B   4) D
5) C   6) A   7) C   8) B
9) A   10) D

Chapter 10: Electrical and RF Safety

1) A   2) D   3) C   4) D
5) B   6) D   7) A   8) B
9) C   10) A

Practice Exam 1

1) A   2) C   3) C   4) D
5) A   6) B   7) C   8) A
9) B   10) D   11) C   12) D
13) D   14) B   15) D   16) C
17) A   18) C   19) C   20) C
21) A   22) B   23) C   24) B
25) D   26) C   27) B   28) B
29) C   30) D   31) B   32) A
33) D   34) D   35) A

Practice Exam 2

1) C   2) A   3) C   4) A
5) C   6) C   7) A   8) D
9) C   10) A   11) D   12) A
13) B   14) C   15) A   16) A
17) D   18) A   19) B   20) A
21) D   22) C   23) D   24) A
25) C   26) B   27) D   28) B
29) A   30) B   31) C   32) D
33) B   34) B   35) B

Practice Exam 3

1) D	2) B	3) C	4) B
5) D	6) D	7) B	8) B
9) D	10) B	11) D	12) B
13) B	14) D	15) B	16) D
17) B	18) B	19) D	20) B
21) B	22) A	23) C	24) C
25) B	26) D	27) C	28) D
29) D	30) A	31) A	32) A
33) C	34) C	35) D	

You're only one step away!

With a General class FCC license, you have access to 83 percent of all amateur radio operating frequencies. Why not unlock the whole 100 percent with an Amateur Extra license?

Less than 20 percent of ham radio operators make it to the prestigious Extra level license, the highest license level available from the FCC.

Here at Ham Radio Prep, we believe that you can reach this master status with a little bit of hard work and studying!

Enroll in our online Extra class course available at: HamRadioPrep.com/amateur-extra-license/

Or check out our printed Extra study book on Amazon!